实时数据处理和
分析指南

[印度] 希尔皮·萨克塞纳（Shilpi Saxena） [印度] 沙鲁巴·古普塔（Saurabh Gupta） 著

吴志国 曾凤姝 译

人民邮电出版社
北京

图书在版编目（CIP）数据

实时数据处理和分析指南 /（印）希尔皮·萨克塞纳
(Shilpi Saxena)，（印）沙鲁巴·古普塔
(Saurabh Gupta) 著；吴志国，曾凤姝译. -- 北京：
人民邮电出版社，2020.5（2021.1重印）
ISBN 978-7-115-52486-7

Ⅰ. ①实… Ⅱ. ①希… ②沙… ③吴… ④曾… Ⅲ.
①数据处理－指南 Ⅳ. ①TP274-62

中国版本图书馆CIP数据核字(2020)第052974号

版 权 声 明

Copyright ©Packt Publishing 2018. First published in the English language under the title
Practical Real-time Data Processing and Analytics (9781787281202).
All rights reserved.

本书由英国 Packt Publishing 公司授权人民邮电出版社有限公司出版。未经出版者书面许可，对本书的任何部分不得以任何方式或任何手段复制和传播。

版权所有，侵权必究。

- ◆ 著 ［印度］希尔皮·萨克塞纳（Shilpi Saxena）
  ［印度］沙鲁巴·古普塔（Saurabh Gupta）
  译 吴志国 曾凤姝
  责任编辑 吴晋瑜
  责任印制 王 郁 焦志炜
- ◆ 人民邮电出版社出版发行　北京市丰台区成寿寺路 11 号
  邮编 100164　电子邮件 315@ptpress.com.cn
  网址 http://www.ptpress.com.cn
  北京天宇星印刷厂印刷
- ◆ 开本：800×1000　1/16
  印张：18.5
  字数：382 千字　　　2020 年 5 月第 1 版
  印数：2 401 – 2 900 册　2021 年 1 月北京第 2 次印刷

著作权合同登记号　图字：01-2018-8913 号

定价：79.00 元
读者服务热线：(010)81055410　印装质量热线：(010)81055316
反盗版热线：(010)81055315
广告经营许可证：京东市监广登字 20170147 号

内容提要

本书主要介绍实时大数据计算领域的相关技巧和经验，包括 Flink、Spark 和 Storm 等流处理框架技术。全书从搭建开发环境开始，逐步实现流处理，循序渐进地引导读者学习如何利用 Rabbit MQ、Kafka 和 NiFi 以及 Storm、Spark、Flink 和 Beam 等组件协同应用来解决实际问题。

本书内容分为 6 个部分，分别是"导言——熟悉实时分析""搭建基础设施""Storm 实时计算""使用 Spark 实现实时计算""使用 Flink 实现实时分析"以及"综合应用"。

在阅读本书之前，读者应具备基本的 Java 和 Scala 编程基础，还应熟悉 Maven、Java 和 Eclipse 的安装和配置流程。

作者简介

希尔皮·萨克塞纳（Shilpi Saxena）是 IT 从业者，也是一位技术布道者。她是一名工程师，曾涉足多个领域（机器对机器空间、医疗保健、电信、人才招聘和制造业）。在企业解决方案的构思和执行的所有方面，她都有着丰富的经验。过去 3 年来，她一直在大数据领域从事设计、管理和提供解决方案的工作。她还负责管理一个分布在世界各地的精英工程师团队。

希尔皮在软件行业的产品和服务方面有超过 12 年（大数据领域 3 年）的开发和执行企业解决方案的经验。她曾担任过开发者、技术负责人、产品负责人、技术经理等职位，可以说在这个行业阅历颇丰。她通过 AWS 的自动扩展，设计并完成了一些在大数据领域中基于 Storm 和 Impala 的前沿的产品实现。

希尔皮参与编写了 *Real-time Analytics with Storm and Cassandra* 一书（Packt 出版社出版）。

沙鲁巴·古普塔（Saurabh Gupta）是一名软件工程师，已有数十年的 IT 行业从业经验，在大数据领域有超过 3 年的工作经验，目前从事处理和设计在生产中运行的实时和批处理项目的相关工作，主要包括 Impala、Storm、NiFi、Kafka 等技术以及在 AWS 上部署 Docker，他还参与了各种物联网项目，涉及电信、医疗保健、智能城市、智能汽车等领域。

前言

本书给出了实时大数据计算领域的许多技巧和经验，介绍了 Flink、Spark 和 Storm 等流处理框架技术。本书还归纳了一些实用的技术，以帮助读者像使用 Hadoop 批处理一样的方式实时处理无界流数据。读者可以从如何搭建开发环境开始，逐步实现流处理，然后学会如何利用 Rabbit MQ、Kafka 和 NiFi 以及 Storm、Spark、Flink 和 Beam 等组件协同应用来解决实际问题。通过学习本书的内容，读者可以对 NRT 的基本原理及应用有透彻的理解，并能掌握如何将这些基础知识应用到任何适用的实际问题当中。

本书采用"菜谱"（Cookbook）式的写作风格，辅以丰富的实际案例，包括注释清楚的代码示例、相应的图表等。

本书内容概述

第一部分　导言——熟悉实时分析　本部分主要带领读者熟悉实时分析领域，了解它的基础组件和基于此构建的系统，包括如下几章：

- 第 1 章　实时分析简介
- 第 2 章　实时应用的基本组件

第二部分　搭建基础设施　本部分主要讲解如何由基础组件搭建基础设施，包括如下几章：

- 第 3 章　了解和跟踪数据流
- 第 4 章　安装和配置 Strom
- 第 5 章　配置 Apache Spark 和 Flink

第三部分　Storm 实时计算　本部分主要关注 Strom 的计算能力和它的各种特性，

包括如下几章：

- 第 6 章　集成 Storm 与数据源
- 第 7 章　从 Storm 到 Sink
- 第 8 章　Storm Trident

第四部分　使用 Spark 实现实时计算　本部分主要关注 Spark 的计算能力和它的相关特性，包括如下几章：

- 第 9 章　运用 Spark 引擎
- 第 10 章　运用 Spark 操作
- 第 11 章　Spark Streaming

第五部分　使用 Flink 实现实时分析　本部分主要关注 Flink 的计算能力和它的相关特性，包括如下一章：

- 第 12 章　运用 Apache Flink

第六部分　综合应用　本部分包括如下一章：

- 第 13 章　用例研究

阅读基础

本书旨在引导读者逐步掌握实时流处理技术。在阅读本书之前，读者应具备基本的 Java 和 Scala 编程基础，还应熟悉 Maven、Java 和 Eclipse 的安装和配置流程，以便运行示例程序。

读者对象

如果读者是 Java 开发人员，想要安装相关软件并设计一个端到端的实时数据流的实用解决方案，那么本书非常适合作为参考书。掌握实时处理的基本知识是很有帮助的，了解 Maven、Shell 和 Eclipse 的基本原理也对读者大有裨益。

本书约定

在本书中，读者会发现许多文本样式，可以据此区分不同种类的信息。下面给出了这些样式的一些例子，并对它们的含义进行了解释。文本中的代码、数据库表名、文件夹名、文件扩展名、路径名、虚拟 URL、用户输入和 Twitter 句柄表示为："下载 `kafka_2.11-0.10.1.1.tgz` 文件后，提取文件。"

代码块设置如下：

```
cp kafka_2.11-0.10.1.1.tgz /home/ubuntu/demo/kafka
cd /home/ubuntu/demo/kafka
tar -xvf kafka_2.11-0.10.1.1.tgz
```

新术语和**重要单词**以粗体显示。读者在截屏图中看到的单词（例如，在菜单或对话框中）在文本中表示为："为了下载新模块，我们将转到 **Files | Settings | Project Name | Project Interpreter**。"

警告或重要注释的形式如下。

 警告内容。

提示和窍门的形式如下。

 提示内容。

审稿人简介

鲁本·奥利瓦·拉莫斯（Ruben Oliva Ramos） 是莱昂技术学院的计算机系统工程师，他毕业于墨西哥瓜纳华托州莱昂市的 Salle Bajio 大学，拥有该校计算机和电子系统工程、远程信息学和网络专业的硕士学位。他在开发 Web 应用程序方面有 5 年以上的经验，擅长用 Web 框架和云服务来控制和监控与 Arduino 和 Raspberry Pi 连接的设备，进而构建物联网应用程序。

鲁本·奥利瓦·拉莫斯在墨西哥的 Salle Bajio 大学的机电一体化系任教，是机电一体化系统设计和工程硕士生导师。他还在墨西哥瓜纳华托州莱昂市的一家机构（Centro de Bachillerato Tecnologico Industrial 225）工作，负责教电子、机器人和控制、自动化和微控制器等课程。他也是一些监控系统和数据记录仪项目的顾问和开发人员——用编程技术（如 Android、iOS、Windows Phone、HTML5、PHP、CSS、AJAX、JavaScript、Augular 和 ASP.NET）、数据库（如 SQlite、MongoDB 和 MySQL）、Web 服务器（如 Node.js 和 IIS）以及硬件编程（如 Arduino、Raspberry Pi、Ethernet Shield、GPS 和 GSM/GPRS、ESP8266）来实现数据采集和编程的控制和监控系统。

他撰写了 *Internet of Things Programming with JavaScript* 一书，该书由 Packt 出版社出版，并参与了用 Arduino 和 Visual Basic .NET 为 Alfaomega 监控、控制和获取数据的项目。

感谢在参与这个项目的过程中给予我帮助和理解的人们，他们是：我亲爱的妻子 Mayte、我两个可爱的儿子 Ruben 和 Dario、我亲爱的父亲 Ruben 和母亲 Rosalia、我的弟弟 Juan Tomas 和妹妹 Rosalia。在我审阅这本书的过程中，他们给了我很多的支持，让我能够追求自己的梦想，并容忍我在忙碌的一天工作后不能陪伴他们。

胡安·汤玛斯·奥利瓦·拉莫斯（Juan Tomás Oliva Ramos）是一名环境工程师，毕业于墨西哥瓜纳华托大学，获得了工程和质量管理的硕士学位。他在专利管理和开发、技术创新项目以及通过控制过程的统计来开发技术解决方案领域有超过 5 年的经验。自 2011 年以来，他一直担任统计、创业和项目技术开发的教师。他还是企业家导师，并在 Instituto Tecnologico Superior de Purisima del Rincon 开设了一个新的技术管理和创业系。

胡安是 Alfaomega 的审稿人，曾参与了 *Wearable designs for Smart watches, Smart TVs and Android mobile devices* 一书的工作。他还通过编程和自动化技术开发了用于改进操作的原型（这些原型已经注册了专利）。

感谢 Packt 让我有机会审校这本令人惊叹的书，并能有幸与一群敢于担当的人合作。

还要感谢我美丽的妻子 Brenda、我的两个女儿 Regina 和 Renata 以及我们家的新成员 Angel Tadeo——感谢你们给了我力量，让我幸福和快乐地度过人生中的每一天。谢谢你们成为我的家人。

普拉蒂克·巴蒂（Prateek Bhati）毕业于印度最为知名的私立大学——阿米提大学。他目前居住在新德里，就职于 Accenture 公司，已有 4 年的实时数据处理经验。

资源与支持

本书由异步社区出品,社区(https://www.epubit.com/)为您提供相关资源和后续服务。

配套资源

本书为读者提供示例源代码。读者可登录异步社区本书页面进行下载。

提交勘误

作者和编辑尽最大努力来确保书中内容的准确性,但难免会存在疏漏。欢迎读者将发现的问题反馈给我们,帮助我们提升图书的质量。

读者如果发现错误,请登录异步社区,按书名搜索,进入本书页面,单击"提交勘误",输入勘误信息,单击"提交"按钮即可。本书的作者和编辑就读者提出的勘误进行审核,确认并接受后,将赠予读者异步社区的 100 积分(积分可用于在异步社区兑换优惠券、样书或奖品)。

扫码关注本书

读者可以扫描下方的二维码,在异步社区微信服务号中看到本书信息及相关的服务提示。

资源与支持

与我们联系

我们的联系邮箱是 contact@epubit.com.cn。

如果读者对本书有任何疑问或建议，请发邮件给我们，并在邮件标题中注明本书书名，以便我们更高效地做出反馈。

如果读者有兴趣出版图书、录制教学视频，或者参与图书翻译、技术审校等工作，可以发邮件给我们；有意出版图书的作者也可以到异步社区在线提交投稿（直接访问 www.epubit.com/selfpublish/submission 即可）。

如果读者来自学校、培训机构或企业，想批量购买本书或异步社区出版的其他图书，也可以发邮件给我们。

如果读者在网上发现有针对异步社区出品图书的各种形式的盗版行为，包括对图书全部或部分内容的非授权传播，请将怀疑有侵权行为的链接发邮件给我们。这既是对作者权益的保护，也是我们提供高品质内容的动力之源。

关于异步社区和异步图书

"异步社区"是人民邮电出版社旗下 IT 专业图书社区，致力于出版精品 IT 技术图书和相关学习产品，为作译者提供优质出版服务。异步社区创办于 2015 年 8 月，提供大量精品 IT 技术图书和电子书，以及高品质技术文章和视频课程。更多详情请访问异步社区官网 https://www.epubit.com。

"异步图书"是由异步社区编辑团队策划出版的精品 IT 专业图书的品牌，依托于人民邮电出版社近 30 年的计算机图书出版积累和专业编辑团队，相关图书在封面上印有异步图书的 Logo。异步图书的出版领域包括软件开发、大数据、AI、测试、前端、网络技术等。

异步社区

微信服务号

目录

第一部分 导言——熟悉实时分析

第1章 实时分析简介 ········· 2
- 1.1 大数据的定义 ········· 2
- 1.2 大数据的基础设施 ········· 3
- 1.3 实时分析——神话与现实 ········· 6
- 1.4 近实时解决方案——可用的架构 ········· 9
  - 1.4.1 NRT 的 Storm 解决方案 ········· 9
  - 1.4.2 NRT 的 Spark 解决方案 ········· 10
- 1.5 Lambda 架构——分析可能性 ········· 11
- 1.6 物联网——想法与可能性 ········· 13
- 1.7 云——考虑 NRT 和物联网 ········· 17
- 1.8 小结 ········· 18

第2章 实时应用的基本组件 ········· 19
- 2.1 NRT 系统及其构建模块 ········· 19
  - 2.1.1 数据采集 ········· 21
  - 2.1.2 流处理 ········· 22
  - 2.1.3 分析层——服务终端用户 ········· 23
- 2.2 NRT 的高级系统视图 ········· 25
- 2.3 NRT 的技术视图 ········· 26
  - 2.3.1 事件生产者 ········· 27
  - 2.3.2 数据收集 ········· 27
  - 2.3.3 代理 ········· 29
  - 2.3.4 转换和处理 ········· 31
  - 2.3.5 存储 ········· 32
- 2.4 小结 ········· 32

第二部分 搭建基础设施

第3章 了解和跟踪数据流 ········· 34
- 3.1 了解数据流 ········· 34
- 3.2 为数据提取安装基础设施 ········· 35
  - 3.2.1 Apache Kafka ········· 35
  - 3.2.2 Apache NiFi ········· 36
  - 3.2.3 Logstash ········· 41

	3.2.4 Fluentd ………… 43	4.5 小结 …………………………… 67
	3.2.5 Flume …………… 46	第5章 配置 Apache Spark 和 Flink … 68
3.3	将数据从源填到处理器—— 期望和注意事项 ………… 48	5.1 安装并快速运行 Spark ……… 68
		5.1.1 源码构建 ………………… 69
3.4	比较与选择适合用例的最佳 实践 ………………………… 49	5.1.2 下载 Spark 安装包 ……… 69
		5.1.3 运行示例 ………………… 70
3.5	小试牛刀 …………………… 49	5.2 安装并快速运行 Flink ……… 73
3.6	小结 ………………………… 51	5.2.1 使用源码构建 Flink …… 73
第4章	安装和配置 Storm …………… 52	5.2.2 下载 Flink ……………… 74
4.1	Storm 概述 ………………… 52	5.2.3 运行示例 ………………… 75
4.2	Storm 架构和组件 ………… 53	5.3 安装并快速运行 Apache Beam ………………… 79
	4.2.1 特征 …………………… 54	
	4.2.2 组件 …………………… 54	5.3.1 Beam 模型 ……………… 79
	4.2.3 流分组 ………………… 56	5.3.2 运行示例 ………………… 80
4.3	安装和配置 Storm ………… 57	5.3.3 MinimalWordCount 示例 ………………………… 82
	4.3.1 安装 Zookeeper ……… 57	
	4.3.2 配置 Apache Storm …… 59	5.4 Apache Beam 中的平衡 …… 85
4.4	在 Storm 上实时处理任务 … 61	5.5 小结 ………………………… 88

第三部分 Storm 实时计算

第6章	集成 Storm 与数据源 ……… 90	6.3 RabbitMQ 与 Storm 集成 …… 99
6.1	RabbitMQ 有效的消息传递 … 90	6.4 PubNub 数据流发布者 …… 107
6.2	RabbitMQ 交换器 ………… 91	6.5 将 Storm 和 RMQ_PubNub 传感器 数据拓扑串在一起 ……… 111
	6.2.1 直接交换器 …………… 91	
	6.2.2 RabbitMQ 安装配置 … 93	6.6 小结 ………………………… 114
	6.2.3 RabbitMQ 的发布和 订阅 ………………………… 95	第7章 从 Storm 到 Sink …………… 115
		7.1 安装并配置 Cassandra …… 115

	7.1.1 安装 Cassandra ………… 116	8.4	Trident 操作 ……………………… 149
	7.1.2 配置 Cassandra ………… 117		8.4.1 函数 ………………………… 149
7.2	Storm 和 Cassandra 拓扑 … 118		8.4.2 Map 函数 and FlatMap
7.3	Storm 和 IMDB 集成处理维度		函数 ………………………………… 150
	数据 ……………………………………… 120		8.4.3 peek 函数 ………………… 151
7.4	集成表示层与 Storm ……… 122		8.4.4 过滤器 ……………………… 151
7.5	小试牛刀 ………………………… 134		8.4.5 窗口操作 …………………… 152
7.6	小结 ……………………………………… 143		8.4.6 聚合操作 …………………… 155
第 8 章	Storm Trident ………………… 144		8.4.7 分组操作 …………………… 158
8.1	状态保持和 Trident ……… 144		8.4.8 合并和组合操作 ……… 159
	8.1.1 事务性 spout …………… 145	8.5	DRPC …………………………………… 160
	8.1.2 不透明事务性 spout … 145	8.6	小试牛刀 ………………………… 161
8.2	基本 Storm Trident 拓扑 … 146	8.7	小结 ……………………………………… 164
8.3	Trident 内部实现 …………… 148		

第四部分 使用 Spark 实现实时计算

第 9 章	运用 Spark 引擎 …………… 166	第 10 章	运用 Spark 操作 …………… 180
9.1	Spark 概述 …………………………… 166	10.1	Spark——封装和 API ……… 180
9.2	Spark 的独特优势 …………… 169	10.2	RDD 语用探索 ………………… 182
9.3	Spark 用例 …………………………… 172		10.2.1 转换 ………………………… 185
9.4	Spark 架构——引擎内部的		10.2.2 动作 ………………………… 190
	运行模式 ………………………… 174	10.3	共享变量——广播变量和
9.5	Spark 的语用概念 …………… 176		累加器 ………………………………… 192
9.6	Spark 2.x——数据框和数据集的		10.3.1 广播变量 ………………… 192
	出现 ……………………………………… 178		10.3.2 累加器 …………………… 195
9.7	小结 ……………………………………… 179	10.4	小结 ……………………………………… 196

目录

第 11 章 Spark Streaming ········ 197
11.1 Spark Streaming 的概念 ········ 197
11.2 Spark Streaming 的简介和体系结构 ········ 198
11.3 Spark Streaming 的封装结构 ········ 203
11.3.1 Spark Streaming API ········ 203
11.3.2 Spark Streaming 操作 ········ 204
11.4 连接 Kafka 和 Spark Streaming ········ 206
11.5 小结 ········ 208

第五部分 使用 Flink 实现实时分析

第 12 章 运用 Apache Flink ········ 210
12.1 Flink 体系结构和执行引擎 ········ 210
12.2 Flink 的基本组件和进程 ········ 213
12.3 将源流集成到 Flink ········ 215
12.3.1 和 Apache Kafka 集成 ········ 215
12.3.2 和 RabbitMQ 集成 ········ 218
12.4 Flink 处理和计算 ········ 221
12.4.1 Datastream API ········ 221
12.4.2 DataSet API ········ 223
12.5 Flink 持久化 ········ 224
12.6 FlinkCEP ········ 226
12.7 Pattern API ········ 227
12.7.1 检测模式 ········ 227
12.7.2 模式选择 ········ 228
12.7.3 示例 ········ 228
12.8 Gelly ········ 229
12.9 小试牛刀 ········ 231
12.10 小结 ········ 242

第六部分 综合应用

第 13 章 用例研究 ········ 244
13.1 概述 ········ 244
13.2 数据建模 ········ 245
13.3 工具和框架 ········ 246
13.4 建立基础设施 ········ 247
13.5 实现用例 ········ 252
13.5.1 构建数据模拟器 ········ 252
13.5.2 Hazelcast 加载器 ········ 259
13.5.3 构建 Storm 拓扑 ········ 261
13.6 运行用例 ········ 272
13.7 小结 ········ 279

第一部分 导言——熟悉实时分析

- 第1章 实时分析简介
- 第2章 实时应用的基本组件

第 1 章
实时分析简介

本章为读者展现了大数据技术的全貌,尤其是大数据实时分析的概况。本章先给出了概念性的大纲,旨在起到抛砖引玉的作用,以激励读者继续阅读本书后续章节的内容。

---本章主要包括以下内容---

- 大数据的定义
- 大数据的基础设施
- 实时分析——神话与现实
- 近实时解决方案——可用的架构
- 分析学——过多的可能性
- 物联网——想法与可能性
- 云——考虑 NRT 和物联网

1.1 大数据的定义

简单来说,大数据有助于处理"3V"问题——体量、速度和多样性。最近,又增加了"2V"——真实性与价值,这就构成了一个五维的范式。

体量:数据的数量。环顾四周,每时每刻都有大量的数据产生,比如电子邮件、推特(Twitter)、脸书(Facebook)或者其他社交媒体中的信息,又如视频、图片、短信、电话记录以及各种设备和传感器产生的数据。数据的计量单位从 TB 级到 ZB 级,甚至到 YB 级这样趋近天文数字的量级。在 Facebook 上,每天大约产生 100 亿条消息,点赞 50

亿次，上传4亿张照片。统计结果令人惊讶，2008年前产生的所有数据量与今天一天生成的数据量相当，相信在不远的将来，这个时间很快就会缩短为一小时。仅从数据体量这一维度来看，传统数据库已经无法在合理的时间范围内存储和处理大规模数据，于是大数据栈脱颖而出，它以低成本、分布式且可靠有效的方式处理这些惊人的海量数据。

速度：数据产生的速度。如今的时代，各种各样的数据都在激增。正是因为数据产生的速度足够快，才积累了如此海量的数据。社交媒体上的事件通常在数秒内就开始流传，接着就开始病毒式地传播。股票交易员在短短数毫秒内就能从社交媒体的热门事件中分析出一些有用信息，并由此触发大量的买入/卖出操作。大数据赋予人们以惊人的速度分析数据的能力：在零售业柜台的终端设备上，短短数秒内信用卡刷卡、欺诈交易的辨别、支付、记账和确认回执等一系列操作就都完成了。

多样性：该维度呈现这样一个事实——大数据很可能是非结构化的。在传统数据库时代甚至更早以前，大部分人习惯于处理类似于表格这样非常结构化的数据。如今超过80%的数据是非结构化的，如照片、短视频、社交媒体更新、传感器采集的数据和通话录音等。大数据技术让你以结构化方式存储和处理非结构化数据，实际上这在一定程度上消除了多样性。

真实性：该维度关乎数据的有效性和准确性。应该如何判断数据是否准确和有效呢？海量的数据记录并非都是经过修正的、准确的且可作为参考的。真实性的内涵在于数据的可信度和质量是怎么样的。数据真实性的例子包括Facebook和Twitter上的帖子使用了不标准的缩写且有拼写错误。大数据已将对数据进行分析的功能用于数据表中。决定数据量究竟有多大的主要因素就是真实性。

价值：顾名思义，就是数据实际拥有的价值。毫无疑问，这是大数据中最重要的维度。从超大型数据集中获取一些有价值的信息或许是人们处理它们的唯一动机，因为所有这些都关乎成本和效益。

当前，几乎所有企业都十分关注大数据技术。众多行业都深信它的实用价值，但实现如上目标的关键主要是面向应用程序，而不是面向基础设施。下一节会详细介绍这部分内容。

1.2 大数据的基础设施

在深入探究大数据基础设施之前，我们先带领读者一览大数据的全貌。表1-1从高层次视角对大数据细分领域进行了划分。

表 1-1

细分领域	典型企业或软件
垂直应用程序	Predictive policing、BloomReach、Myrrix
广告、媒体应用	Media Science、Turn、Recorded Future
数据即服务	Factual、Gnip、Kaggle
商业智能	Oracle、SAP、IBM
日志数据应用程序	Splunk、Loggly、Sumo Logic
数据分析基础设施	Hortonworks、Cloudera、DataStax
可运维基础设施	Couchbase、Teradata、Hadapt
基础设施即服务	亚马逊网络服务、Microsoft Azure、Google 云平台
核心技术	Apache Hadoop、Apache HBase、Apache Cassandra
结构化数据库	Microsoft SQL Server、MySQL、PostgreSQL

表 1-1 描述了大数据技术的细分领域。最底层是最关键的，支持可扩展和分布式存储。

- **核心技术**：核心底层软件如 Hadoop、MapReduce、Mahout、HBase 和 Cassandra 等。再往上一层是基础设施层，开发者依据用例和解决方案可以选择合适的基础设施。

- **数据分析基础设施**：EMC、Netezza、Vertica、Cloudera 和 Hortonworks。

- **可运维基础设施**：Couchbase、Teradata 和 Informatica 等。

- **基础设施即服务**（IaaS）：AWS 和 Google 云等。

- **结构化数据库**：Oracle、SQLServer、MySQL 和 Sybase 等。往上一层是特定需求的**数据即服务**（DaaS）：Kaggle、Azure 和 Factual 等。

- **商业智能**（BI）：Qlikview、Cognos 和 SAP BO 等。

- **分析和可视化**：Pentaho、Tableau 和 Tibco 等。

可以看到，如今传统的关系型数据库仍然在为数据存储及处理实现高效和低成本的效果而努力挣扎。传统的关系型数据库处理大数据的成本非常高，通过扩展的方式很难满足低延迟的要求。正是由于以上现状,才促进了具有低成本、低时延、高扩展性、开源等需求的新技术的涌现。黄色的大象——Hadoop 成为救星,它出其不意地占领了数据存储和计算的竞技场。Hadoop 作为分布式数据存储和计算框架,在设计上具有非常高的可靠性和可扩展性。Hadoop 计算方法的核心是将数据分块存储在集群的所有节点上,然后在所有节点上并行地处理数据。

相信到了这里,读者已经对大数据的基础知识和全貌有了一些认识,能够以 Hadoop 框架为例来深入研究大数据的概念。接下来继续研究实现 Hadoop 集群的体系结构和方法,这与高层基础设施和大型数据集群的典型存储需求非常相似。本书将深入研究的另一个关键话题是大数据环境下的信息安全。图 1-1 主要指出大数据基础设施中的几个关键因素。

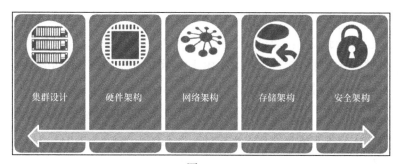

图 1-1

集群设计:这是基础设施规划中最重要且最有决定性的一个因素。基础设施的集群设计策略基本上是解决方案的主要考虑因素,包括应用程序用例和要求、工作负载、资源计算(取决于是内存密集型还是计算密集型)以及安全性。除了计算、内存和网络利用率,另一个重要因素是存储,它将基于云或本地。云的选择有公共云、私有云或混合云,这取决于应用场景和企业的需求。

硬件架构:存储成本主要取决于存储数据的体量、存档策略以及数据生存期限。决定性因素有两点。第一点是实现的算力需求(商用化组件是否丰富,或者是否需要高性能 GPU)。第二点,内存需求是什么?是低等、中等,还是高等?这取决于应用程序实现内存算力需求。

网络架构:这听起来可能不是很重要,但它是大数据应用的一个重要驱动力。原因在于大数据的关键是分布式计算,而且网络利用率比单服务器单片集成实现的情况高得多。在分布式计算中,数据负载和中间计算结果在网络上传输。因此,网络带宽成为总体解决方案的节流代理,并且取决于基础设施策略的主要方面的选择。糟糕的设计方法有时会导致网络阻塞,其中数据在处理上花费的时间更少,而在通过网络或等待传输以供下一步执行所花费的时间更多。

安全架构:安全对于任何应用程序来说都是非常重要的。在大数据应用场景下,由于它的体量和多样性,以及计算需要通过网络获取数据,因此安全就变得更加重要。安全对大数据基础设施具有关键性和战略性意义,云计算和存储选型这两方面进一步增加了未来对其需求的复杂性。

1.3 实时分析——神话与现实

实时分析的最大真相是实际上没有什么东西是真正实时的,这仅仅是一个神话。实际上,只能说它接近于实时。通过分析可以得到这样的结论:只有提高解决方案的性能和减少操作延时,分析才能接近于实时。由于实际中计算、操作和网络的延迟,实际上不可能消除实时和近实时之间的差距。

在进一步讨论之前,我们带领读者快速了解一下这些所谓的实时分析解决方案的高层次需求。图 1-2 展现了满足高层次需求的一个系统,该系统可以使用各种结构化和非结构化数据集处理数百万个事务。首先,程序引擎应该超快,并能够处理非常复杂的连接操作和多样化的业务逻辑;其次,可以准确产生令人叹为观止的报告,在一瞬间恢复即席查询(AdHoc 查询),并在没有延迟的情况下渲染可视化的仪表面板。

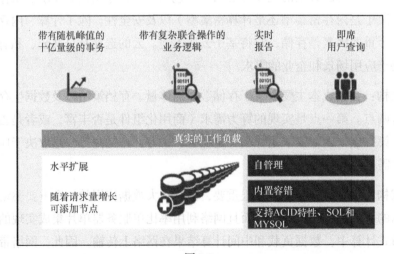

图 1-2

以前对实时解决方案的要求是不够的，如果把它们推广到生产环境中，即在当今的数据生成和零停机时代，最基本的要求是，系统应该能够以最小的代价实现自我管理或被管理，并且以容错和自动恢复的方式来构建，以处理大多数情况（即便不是所有情况）。它还应该能够提供类似于基本 SQL 的接口。

尽管前面对实时分析的要求听起来有些极端可笑，但是它们都是当今大数据解决方案最正常和最基本的要求。然而，回到实时分析这个主题，既然已经简要地谈到了数据、处理和输出方面的系统级要求，这些正在设计和已被设计的系统用于处理数以万计的事务并动态应用复杂的数据科学和机器学习算法，以尽可能接近实时地计算结果。图 1-3 描述了计算时间、上下文的概念以及最终见解的重要意义。

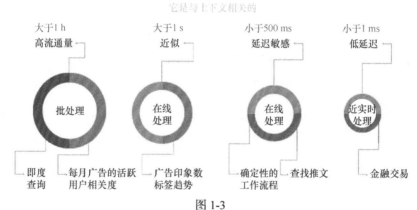

图 1-3

如图 1-3 所示，在有限时间背景下，存在以下问题。

- 对泽字节（ZB）数据的即席查询占用了小时级的时间，因此这通常被称为批处理。图 1-3 中圆的大小比喻的是以图形式处理的数据的大小。
- 广告展示次数/标签广告趋势/确定性工作流程/推文：这些大多被称为在线时间和计算时间的用例通常为 500ms/1s。虽然与以前的用例相比，计算时间大大减少了，但是处理的数据量也显著减少了。它可以非常迅速处理几吉字节大小（GB）的数据流。
- 财务跟踪/关键任务应用程序：典型特点是数据量很低，数据率非常高，处理非常快，并且在几毫秒的时间窗口中产生低延迟计算结果。

除了计算时间，批处理、实时处理以及解决方案设计之间还有一些显著的差异，见表1-2。

表 1-2

批处理	实时处理
静态数据	动态数据
批大小有界	数据以流的形式存在，是无界的
访问全部数据	访问当前事务/滑动窗口内的数据
数据以批的形式处理	数据在事件、窗口或者微批级别上处理
高效、易于管理	实时分析，但是系统相对于批处理较为脆弱

在本节，我们想强调的是**近实时**（NRT）解决方案是接近真正实时的，因为它实际上是可能实现的。所以，如上所述，RT 实际上是一个神话（或假设），而 NRT 是一个现实。每天处理和查看的 NRT 应用程序，包括车联网、预测和推荐引擎、医疗保健和可穿戴设备。

有一些关键的环节实际上会引入延迟到总周转时间，或者称之为 TAT。实际上，事件发生与产生可行的措施之间的时间间隔是由它产生的。

数据/事件通常通过有线（互联网/电信信道）从不同的地理位置传输到处理中心。这项活动已经过了一段时间。其处理如下。

- **数据着陆**：由于安全方面的原因，数据通常落在边缘节点上，然后被提取到集群中。
- **数据清理**：需要满足数据准确性方面的要求，在处理之前消除错误/不正确的数据。
- **数据修改和丰富**：使用维度数据来绑定和丰富交易数据。
- **实际处理**。
- **存储结果**。这里，所有以前的处理过程都会产生：CPU 周期、磁盘 I/O、网络 I/O、数据序列化方面的主动编组和解组。

既然已经了解了实时分析的实际情况，接下来我们将更深入地了解这些解决方案的架构。

1.4 近实时解决方案——可用的架构

在本节，读者将学会如何构建可扩展、可持续且具有鲁棒性的实时系统解决方案，以及如何对可能的架构模式进行选型。

高级 NRT 解决方案看起来非常直观和简单，其具有数据收集漏斗、分布式处理引擎以及一些其他组件（如缓存、稳定存储和仪表板插件）。

如图 1-4 所示，在较高层面上，基本的分析过程可以分为 3 类：流数据的实时数据收集；分布式流数据的高性能计算；以可查询消耗层/仪表板的形式探索和可视化生成的见解。

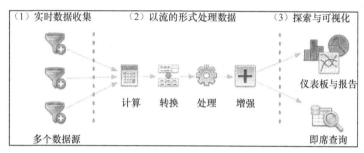

图 1-4

市场上有两种存在竞争的流式计算技术，即 Storm 和 Spark。下面我们将深入研究从这些堆栈中获得的高级 NRT 解决方案。

1.4.1 NRT 的 Storm 解决方案

该解决方案实时捕获高级流数据并将其路由到某个队列/代理（Kafka 或 RabbitMQ）中，然后通过 Storm 拓扑处理分布式处理部分，一旦计算出见解，就可以将它们快速写入数据存储（如 Cassandra）或其他队列（如 Kafka），以进行进一步的实时下游处理。如图 1-5 所示，通过发送/提取收集代理（如 Flume、Logstash、FluentD 或 Kafka 适配器），可从不同数据源收集实时流数据。然后，数据被写入 Kafka 分区，Storm 拓扑从 Kafka 中提取/读取流数据并在其拓扑中处理此数据，并将见解/结果写入 Cassandra 或其他一些实时仪表板。

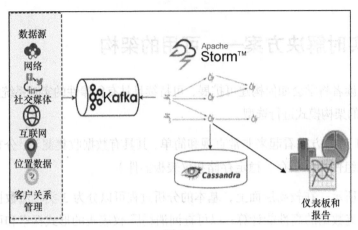

图 1-5

1.4.2 NRT 的 Spark 解决方案

在更高的层级上，Spark 的数据流管道与图 1-5 所示的 Storm 架构非常相似，但是它最受诟病的一点是 Spark 利用 HDFS 作为分布式存储层。在进一步深入之前，我们先看看对整体流程及其细节的进一步剖析，如图 1-6 所示。

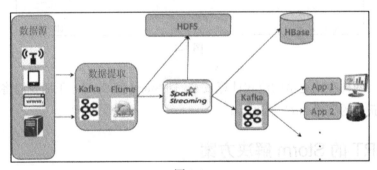

图 1-6

与典型的实时分析管道一样，流数据使用 Flume 或 Logstash 等抓取代理来提取数据。本节首先介绍 Kafka，以确保数据源与抓取代理之间的系统解耦；然后介绍 Spark Streaming 组件——它将结果转储到稳定的存储单元、仪表板或 Kafka 队列之前，提供了一个用于处理数据的分布式计算平台。

前两种架构范式之间有一个本质区别：虽然 Storm 本质上是一个实时事务处理引擎，默认情况下，擅长按事件处理传入数据；但 Spark 基于微批的工作理念，本质上是一个伪实时计算引擎，通过减少微批的大小，可以满足接近实时的期望计算。Storm 主要用于快速处理，所以所有转换都在内存中，因为任何磁盘操作都会产生延迟；对于 Storm 来说，这既是一个优点，又是一个缺点（因为如果事情中断，内存是不稳定的，一切都必须重新处理，中间结果会丢失）。此外，Spark 基本上由 HDFS 支持，并且功能强大且容错性更强，因为中间结果在 HDFS 中有备份。

在过去的几年中，大数据应用程序按以下顺序进行了精彩的转换。

- 仅批处理应用程序（早期的 Hadoop 实现）；
- 仅流处理（早期的 Storm 实现）；
- 可以是两者（前两者的定制组合）；
- 前两者（Lambda 架构）。

问题是：为什么发生了上面的演变？当人们熟悉了 Hadoop 的强大功能时，他们真的很喜欢构建几乎可以处理任何数据量的应用程序，并且可以将其以无缝、容错、无中断的方式扩展到任何级别。然后，随着 Storm 等分布式处理引擎的出现，逐步进入到了一个大数据处理成为强烈需求的时代。Storm 可扩展性强，且具有轻量级快速处理能力。但是，有些情况发生了变化，大部分人意识到了 Hadoop 批处理系统和 Storm 实时系统的局限性和优势：前者满足了对数据量的需求，后者在速度方面非常出色。这些实时应用程序非常完美，它们在整个数据集的短时窗口上表现得很好，但在以后的某个时间没有任何修正数据/结果的机制。虽然 Hadoop 实现准确而强大，但需要花费很长时间才能获得确定性的结论。我们达到了这样的一个程度，即复制了完整/部分解决方案，以获得涉及批处理和实时实现相结合的解决方案。最近的 NRT 架构模式中的 Lambda 架构，是颇受欢迎的解决方案，结合了批处理和实时实现，无须复制和维护两个解决方案。Lambda 架构能同时满足数据量和速度的要求，这是早期架构的优势，可以满足更广泛的用例集。

1.5 Lambda 架构——分析可能性

前文已经介绍了这种神奇的架构，那么本节就仔细研究一下该架构模式。

在底层上，Hadoop 提供了大量存储，并且有 HDFS 和 MapReduce 这种形式的非常强大的处理引擎，既可以处理大量数据又可以执行种种计算。但是，它有很长的**周转时间**（TAT），而且是一个批处理系统，从而可以帮助解决大数据的体量问题。如果对处理速度有要求，需要寻找一种低延迟的解决方案，则必须求助于实时处理引擎，它可以快速处理最新或最近的数据，并且可在有效的时间范围内快速生成一些见解。但是除了速度和快速的 TAT，还需要将更新的数据逐步集成到批处理系统中，以便对整个数据集执行深度批处理分析。因此，从本质上讲，所处的环境既需要批处理系统，也需要实时系统，这种模式的最佳体系结构组合称为 Lambda 架构（λ）。图 1-7 描述了这种模式的高层次设计逻辑。

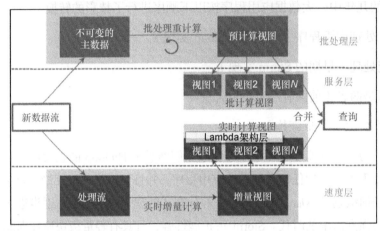

图 1-7

解决方案既与技术无关，又与语言无关；它可以抽象为以下 3 层：批处理层、速度层和服务层。

输入数据被输送到批处理层和速度层，其中批处理层用于创建整个不可变主数据的预计算视图。该层主要有不可变的数据存储，具有一次写入和大量读取的功能。

速度层处理最近的数据，仅维护最近一组数据的增量视图。该层在数据可访问性方面具有随机读取和写入的功能。

问题的症结在于服务层的智能。在服务层中，来自批处理层和速度层的数据被合并，并满足查询需求，因此，可以无缝地从这两者中得到最好的结果。近实时请求是速度层的增量视图（它们具有低保留策略）中的数据来处理的，而引用旧数据的查询是由批处理层中生成的主数据视图来处理的。该层仅适用于随机读取而不能随机写入，但它确以查询、连接以及批量写入的形式来处理批量计算。

但是，Lambda 架构并非是针对所有混合用例的一站式解决方案。有一些关键方面需要注意：总是认为分布式；面向故障的设计和规划；经验法则为数据是不可变的；面向故障设计。

既然已经熟悉了实时分析中流行的架构模式,那么来谈谈这一部分可能存在的用例。图 1-8 展示了可能应用到的高级领域和各种关键用例。

电信	车辆互联	医疗	执法	市场数据分析	物联网
• CDR处理 • 社会分析 • 流失预测 • 地理映射	• 遥感 • 交通管理 • 报警和地理围栏	• ICU检测 • 早期流行病预警 • 远程健康监测	• 多模式实时监控 • 网络安全	• 市场数据分析 • 欺诈检测 • 推荐	• 智能应用互联 • 车辆互联 • 可穿戴设备互联

图 1-8

1.6 物联网——想法与可能性

Kevin Ashton 于 1999 年创造了"物联网"这个术语,并由此成为近 10 年来最有影响力的开拓者之一。虽然有 M2M 形式的物联网前驱和工业自动化仪表控制,但物联网和连接智能设备的时代已经到来,这是前所未有的事情。图 1-9 从俯瞰视角帮助读者了解物联网应用的广泛性和多样性。

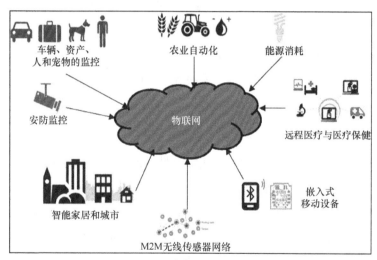

图 1-9

智能互联设备已经进入千家万户,它们具备感知、处理和传输的功能,甚至能够根据处理结果采取行动。几年前科幻小说中的机器时代已经走入现实。如果车主拿着钥匙走进或者远离联网的车,就可以方便地对车进行解锁/锁定。超市里有近距离感应信标,它能感应到顾客和货架的距离,并把报价传送并显示到顾客的手机上。智能办公室通过在空荡荡的会议室里关掉电灯和交流电源来节约电力。这样的例子数不胜数,而且时刻都在增加。

物联网的核心是由联网设备组成的生态系统，这些设备能够在互联网上进行通信。在这里，设备可以是任何东西，像传感器设备、拥有可穿戴设备的人、一个地方、一棵植物、一只动物或一台机器。时至今日，几乎任何我们在这个星球上能想到的实体都可以连接起来。任何物联网平台都主要有 7 层，如图 1-10 所示。

图 1-10

以下是对物联网所有 7 个应用程序层的概述。

- **第 1 层**：设备、传感器、控制器等。
- **第 2 层**：通信信道、网络协议和网络要素、通信、路由硬件-电信、Wi-Fi 和卫星。
- **第 3 层**：基础设施——可以是内部的，也可以是云端的（公共的、私有的或混合的）。
- **第 4 层**：这里是大数据提取层。这是一个平台，收集来自实体/设备的数据，为下一步做准备。
- **第 5 层**：使用复杂处理、机器学习、人工智能等对数据进行清理、解析，消息生成和分析的处理引擎，以报告、警报和通知的形式生成见解。
- **第 6 层**：自定义应用程序、可插拔的二级接口（如可视化仪表板、下游应用程序等）构成这一层的一部分。
- **第 7 层**：这一层中的人员和流程实际上是根据以下系统的建议进行操作的。

在图 1-11 中，如果从自下而上开始，最底层是设备层，它们是传感器或由 RaspberryPi、Ardunio 等计算单元驱动的传感器。此时，通信和数据传输通常由轻量级选项控制，如**消息队列遥测传输（MQTT）**和**约束应用程序协议（CoAP）**，它们正在快速取代 HTTP 等传统选项。该层实际上与聚合或总线层结合在一起，本质上它是一个 Mosquitto 代理，该层从数据源构建了事件传输层，即为从设备到处理集线器之间的部分。一旦到达处理集线器，就可以将计算引擎上的所有数据准备好进行操作，分析和处理数据以生成有用的可操作的命令。这些命令进一步集成到网络服务 API 可消耗层，以用于下游应用程序。除了这些水平层，还有交叉层，它们用于处理设备配置、设备管理、身份和访问管理层。

现在读者了解了标准物联网应用程序的高级架构和层次，下一步是了解物联网解决方案受到限制的关键方面以及对整体解决方案的影响。

安全性：这是整个数据驱动解决方案领域中关键部分之一，连接到互联网的大数据和设备的理念使整个系统更容易受到黑客攻击且安全性方面更敏感，因此在为静态数据和动态数据设计所有层的解决方案时，要将其作为一个战略关注领域来处理。

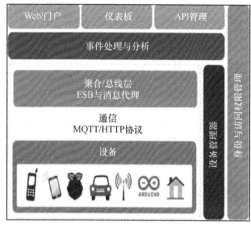

图 1-11

功耗/电池寿命：由于是在为设备而不是人类设计解决方案，因此，应该具有非常低的功耗，且不会消耗电池寿命。

连通性和通信：与人类不同，这些设备总是相互连接，而且非常"健谈"。同样，我们在整体通信方面需要轻量级协议来实现低延迟数据传输。

从故障中恢复：这些解决方案处理数十亿的数据并且维持 7×24 小时的工作模式。该解决方案应该能够诊断故障、应用背压，然后从最小的数据丢失情况中进行自我恢复。如今，物联网解决方案旨在通过检测延迟/瓶颈并具有弹性自动扩展和缩小的能力来处理突然出现的数据峰值。

可扩展性：这个解决方案需要设计为线性可扩展模式，而无须重新构建基础框架或设计，这是因为该域正在使用前所未有且不可预测的设备数量进行扩张，这些设备与全部未来等待发生的用例相连。

接下来是物联网应用框架中先前约束的含义，其表面形式为通信信道、通信协议和处理适配器。

在通信信道供应商方面，物联网生态系统正在从电信信道和 LTE 演变为以下选项。它们分别是：直接以太网/Wi-Fi/3G、**LoRA**、蓝牙低能量（BLE）、**RFID/近场通信**（NFC）、中程无线网状网络（如 Zigbee）。

对于通信协议，事实上的机载标准是 MQTT，其广泛使用的原因是显而易见的。

- 非常轻。
- 就网络利用率而言，占用的空间很小，因此通信速度非常快，负担也更少。

- 提供了一个有保证的传输机制，从而最终能够传输数据，即使是在脆弱的网络上也是如此。
- 功耗低。
- 对网络上的数据包流进行优化，以实现低延迟和较小的使用空间。
- 是一种双向协议，因此既适合从设备上传输数据，也适合将数据传输到设备。
- 更适合这样一种情况，即必须通过网络传输大量的短消息。

边缘分析

后进化、物联网革命和边缘分析是改变游戏规则的重要组成部分。如果要查看物联网应用程序，则需对来自传感器和设备上的数据进行整理，并将其传输到分布式处理单元中，该单元要么位于办公场所，要么位于云上。数据提升和转移导致了大量的网络开销，这使得整个解决方案存在潜在的传输延迟。这些因素催生了一种新的解决方案，并开拓了物联网计算的新领域——边缘分析。顾名思义，它将处理推向边缘，以便数据在其源处被处理。如图1-12所示，物联网分为边缘分析和核心分析。可以看到，物联网的计算现分为以下几个部分。

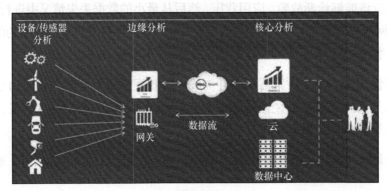

图 1-12

- **设备传感器边缘分析**：其中处理的数据和一些洞见是在设备级得到的。
- **边缘分析**：这些是已处理数据的分析，并在网关级别获得洞见。
- **核心分析**：要求所有数据到达一个公共计算引擎（分布式存储和分布式计算），然后进行高复杂度的处理，以生成可行的洞见，服务于人或机器决策。

传感器/边缘分析的一些典型用例如下。

- **工业物联网（IIoT）**：传感器被嵌入在各种设备、机器，有时甚至是车间里。传感器产生数据，设备本身具有处理数据和产生警报/建议以提高性能或产量的能力。
- **医疗领域的物联网**：智能设备可以参与边缘处理，有助于发现早期预警信号，并对适当的医疗情况提出警报。
- 在可穿戴设备的世界里，边缘处理可以使跟踪和安全变得很容易。

如今，环顾四周，你会发现联网设备在生活、工作中无所不在，如智能交流、智能冰箱和智能电视。这些智能设备都把数据发往中央集线器或者手机上（在那里它们易于控制）。实际上，物联网正在变得越来越智能，正在从互联发展到足以执行计算、处理和预测的智能化程度，例如，咖啡机智能到可以连接到主人的汽车、办公室，能推测主人的日程和到达时间，并随时准备新鲜的热咖啡。

1.7 云——考虑 NRT 和物联网

"云"不过是一个术语，用来识别互联网上可以获得的计算能力。大部分人都熟悉物理机器、服务器和数据中心。云的出现把我们带到了一个虚拟化的世界，在那里我们正在向虚拟节点、虚拟化集群甚至虚拟数据中心转移。现在，使用硬件虚拟化手段在几台物理机上就可以搭建一个虚拟机集群。这就像是让软件运行在硬件上一样。下一步是实现云服务，我们在其上托管了所有虚拟主机上的计算资源，并且可以通过互联网获取。

云服务包括基础设施即服务（Infrastructure as a Service，IaaS）、平台即服务（Platform as a Service，PaaS）和软件即服务（Software as a Service，SaaS）这 3 种类型。

- **基础设施即服务**：它基本上是基于物理计算机的云变体。实际上它通过在网络上运行的虚拟化层取代了实际机器、服务器和硬件存储以及网络。IaaS 允许读者构建整个虚拟基础设施，其实质是模拟实际硬件的软件。
- **平台即服务**：一旦解决了硬件虚拟化部分，下一个显而易见的步骤是考虑在原始计算机硬件上操作的下一层。这是一个将程序和组件绑定在一起的组件，如数据库、服务器、文件存储等。例如，在这里，如果数据库作为 PaaS 公开，那么程序员可以将其用作服务，而不必担心存储容量、数据保护、加密、复制等较低级的细节。PaaS 的著名例子是 Google App 引擎和 Heroku。

第 1 章　实时分析简介

- **软件即服务**：这一层是云计算栈中的最上层，实际上是提供解决方案作为服务的层。这些服务是按每个用户或每月来收费的，这种模式确保最终用户可以灵活地注册和使用服务，而无须支付任何许可证费用或锁定时间。一些广为人知的典型例子是 Salesforce Customer 360 平台和 Google App。

现在我们已经了解并熟悉了云，接下来需要理解的是云计算究竟意味着什么，为什么说"云的出现正在拉下传统数据中心时代的帷幕"。再来了解一下云计算的一些关键优点——这实际上使这个平台成为 NRT 和物联网应用程序的核心。

云服务是按需的。用户可以根据需要和负载提供计算组件/资源。在未来的若干年里，我们没有必要在基础设施上进行巨额投资，也没有必要进行规模化投资，而是可以提供一个足以满足当前需求的集群，然后在需要时通过请求更多的随需应变实例来扩展集群。因此，人作为用户所得到的保证是，在需要一个实例时，会得到一个相同的实例。

云服务允许构建真正有弹性的应用程序。这意味着根据负载和需求，部署可以扩容和降容。这是一个巨大的优势，而且基于云的方式有着很高的成本效益。如果用户有一个应用程序在每个月的第一天流量出现偶发性激增，那么，在云环境下，用户就不需要在 30 天内都提供满足第一天流量激增需求所要配备的硬件。相反，用户可以提供平均一天所需的资源，并构建一种机制来扩展自己的集群，以满足第一天的激增，然后在每月的第二天自动缩至平均大小。

这就是回报。这是云最有趣的特点，它击败了传统硬件供应系统——建立一个数据中心时，必须预先规划金额巨大的投资。在云数据中心环境下，用户不需要这样的成本，只为正在运行的实例付费就够了，而这种付费通常是按小时计算的。

1.8　小结

在本章，我们主要概述了大数据技术的整体面貌，以及大数据作为基础设施和大数据分析的前提意义。我们向读者介绍了在设计和决定大数据基础设施时需要考虑的各种因素和注意事项，还揭开了实时分析和 NRT 架构的真实面纱，并给出了一些可以利用物联网和 NRT 解决问题的案例。在本章的最后，我们简要介绍了物联网的边缘计算和云基础设施的概念。

在第 2 章，我们将带领读者更深入地了解实时分析应用程序、概念与架构，讨论 NRT 应用程序的基本构建模块、所需的技术栈以及开发时可能遇到的挑战。

第 2 章
实时应用的基本组件

本章将带领读者熟悉**近实时**系统的基本构建模块,向读者介绍这些应用的高级逻辑、物理原理和技术视图,并将涉及系统中每个构建模块的技术选型。

---- 本章主要包括以下内容 ----

- NRT 系统及其构建模块
- 数据采集
- 流处理
- 分析层——服务终端用户
- NRT 的高级系统视图
- NRT 的技术视图

2.1 NRT 系统及其构建模块

本章读者遇到的首要问题可能是"什么时候应该将应用程序称为 NRT 应用程序?"简单来讲,一个能够非常接近实时地进行消费、处理和生成结果的应用程序可以称为 NRT 应用程序,也就是说,从事件发生到结果产生的时间间隔非常小,量级从几纳秒到最多几秒。

传统的单体应用无法满足 NRT 应用系统的需求,原因主要有以下几个关键点。

- **后端数据库**:单点单体数据访问。
- **摄入流**:管道复杂,容易导致端到端流的延迟。
- **故障与恢复**:系统容易发生故障,恢复方法困难、复杂。

- **同步和状态捕获**：捕获和维护系统中的事务状态非常困难。多样化的分布式系统和实时系统故障的出现使得这类系统的设计和维护变得更加复杂。

解决上述问题的方案之一是以流式传输为架构，从而基于源源不断的实时数据流使终端用户能够实时看到一些实用的结论。设计流处理系统需要考虑几个挑战，并在以下几点中加以说明。

- 大规模高速系统的局部状态和系统的一致性。
- 数据不会按相同时间间隔到达，它不断流入，并且一直在流式传输。
- 后端数据库没有单一的事实状态，而是应用程序订阅或利用事实数据流。

在进一步深入研究之前，我们有必要先来了解一下时间表示法，如图 2-1 所示。

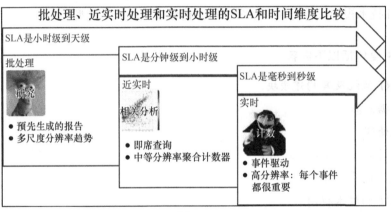

图 2-1

图 2-1 已经非常清楚地将 SLA 与每种实现类型（批处理、近实时和实时）以及每个实现所满足的用例类型关联起来。例如，批处理实现的 SLA 范围从几小时到几天不等，这些解决方案主要用于封闭/预生成报告和趋势预测。实时解决方案具有几秒到几小时的 SLA 量级，可满足需要 AdHoc 查询、中等分辨率的聚合器等场景。在 SLA 方面，实时应用程序是最关键的任务，分辨率是每个事件的计数，其结果必须在事件发生后的毫秒到秒量级延迟后返回。

了解了 NRT、实时和批处理系统的时间维度和 SLA 后，接下来我们讨论了解 NRT 系统的构建模块。

本质上，它由消息传输管道、流处理组件、低延迟数据存储以及可视化和分析工具这 4 个主要组件/层组成，如图 2-2 所示。

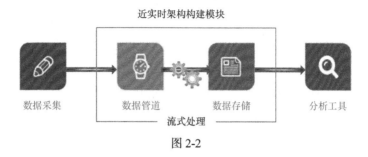

图 2-2

先从源中收集数据并将其提供给**数据管道**，实际上这是一个逻辑管道，它从不同的生产者收集连续事件或流数据并将其提供给消费者的流处理应用程序；然后这些应用程序对这些实时流数据进行转换、整理、关联、聚合和执行各种其他操作，再将结果存储到低延迟数据存储中；最后各种分析、商业智能、可视化工具和仪表板从数据存储中读取这些数据，并将其呈现给商业用户。

2.1.1 数据采集

这是所有数据处理过程的开始。无论是批处理还是实时处理，最主要的挑战是将数据从源处获取到系统中以进行处理。读者可以将处理单元视为一个黑盒和一个数据源，并将消费者视为发布者和订阅者。数据采集的整个过程如图 2-3 所示。

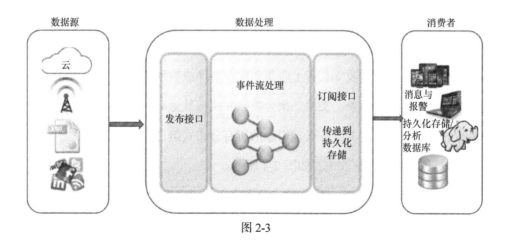

图 2-3

在大数据和实时应用的背景下，衡量数据收集工具的关键指标有：性能和低延迟、可扩展性以及处理结构化和非结构化数据的能力。

除此之外,任何数据收集工具都应该能够接收各种来源的数据,示例如下。

(1)**来自传统跨国系统的数据**:在考虑软件应用时,了解到业内长期以来一直应用传统数据库收集和整理数据。这些数据可能是磁带、Oracle、Teradata、Netezza 等上的顺序文件。因此,从实时应用程序及其相关数据收集开始,系统架构有如下 3 个选项。

- 复制这些传统系统中的 ETL 过程并从源处获取数据。
- 从这些 ETL 系统中提取数据。
- 第三种更好的方法是使用虚拟数据池体系结构复制数据。

(2)**来自物联网/传感器/设备或 CDR 的结构化数据**:这是以非常高的速度和固定格式传输的数据,数据来自各种传感器和电信设备。数据收集/接收的主要复杂性或挑战是数据的多样性和到达速度。收集工具应该能够处理多样性和速度方面的问题,但是对于上游处理来说,这种数据的一个优点是格式非常标准和固定。

(3)**来自媒体文件、文本数据、社交媒体等的非结构化数据**:这是所有传入数据中最复杂的一种,其复杂性是由体量、速度、多样性和结构等方面决定的。数据格式可能差异很大,可能是非文本格式,如音频/视频等。数据收集工具应该能够收集这些数据并将其同化以进行处理。

2.1.2 流处理

流处理组件本身包含如下 3 个子组件。

- **代理程序**:从数据收集代理上收集和保存事件或数据流。
- **处理引擎**:实际转换、关联、聚合数据并执行其他必要操作。
- **分布式缓存**:实际上是在**处理引擎**的所有分布式组件中维护公共数据集的机制。

流处理组件的相似点被放大,如图 2-4 所示。

流处理组件应该满足几个关键属性。分布式组件可以提供故障恢复能力;可扩展性可以满足不断增长的应用程序需求或突发的流量激增;处理此类应用程序预期的总体 SLA 的低延迟;易于操作的用例可以支持不断发展的用例;为故障而构建,系统能够从不可避免的故障中恢复,而不会造成任何事件损失,并且能够从故障点处重新开始处理;

易于集成堆外/分布式缓存或数据存储；各种各样的操作、扩展和功能可以满足用例的商业需求。

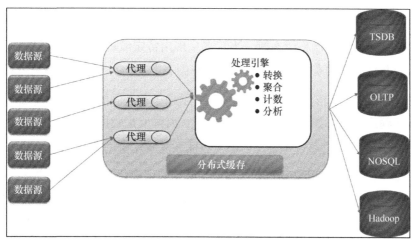

图 2-4

在为流处理应用程序/框架选型时，基本上可以参考以上这些属性。

2.1.3 分析层——服务终端用户

分析层是 NRT 应用程序所有组件中最有创意，也是最为有趣的一部分。截止目前，我们所讨论的都是后端处理环节，但是本节所要介绍的组件将以可视化以及易操作的形式向终端用户展示输出/见解。

作为解决方案的一部分，数据可视化技术选型中最关键的一点是，如何以一种使预期目标受众容易理解并基于此做出某种行动的方式来呈现信息。仅仅对数据进行智能处理并获得可行的见解是不够的，必须接触到参与者，无论他们是人类还是一些流程。

在深入研究商业智能的本质和可视化组件的细节之前，不妨先了解一下大数据和高速 NRT/RT 应用程序给问题混合带来的挑战。

这些可视化系统能够应对的一些挑战如下。

- **对速度的需求**：世界在不断发展和合理化，现在越来越多的公司倾向于实时分析，以在竞争中获得优势。因此，可视化工具应该能补充其所属应用程序的速度，以便能够快速地以有意义的形式描述和呈现关键参与者的准确事实，从而做出明智的决策。

- **理解数据并在正确的背景下呈现**：有时，相同的结果需要进行不同的调整和呈现，这取决于所迎合的受众。因此，该工具应该提供这种灵活性，并能够以最有意义的形式围绕可操作见解设计一个可视化解决方案。例如，如果读者正在绘制城市中的车辆位置，那么可能需要使用热力图，其中颜色差异表示车辆的浓度/数量，而不是在图上绘制每辆车辆。当读者呈现出特定航班/船舶的行程时，可能不希望合并任何数据点，并将它们都绘制在图表上。
- **处理离群值**：图形数据表可以轻松地表示趋势和离群值，从而使终端用户能够发现问题或找出需要注意的点。一般来说，离群值是数据的 1%~5%，但在处理大量数据时，5%的总数据是巨大的，可能会导致绘图问题。

图 2-5 描述了整个应用程序的流程和一些流行的可视化方法，包括 Twitter 热力图。

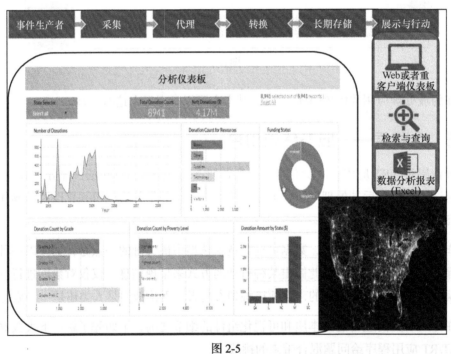

图 2-5

该图描述了从事件生产者到收集代理的信息流，紧接着是代理和处理引擎（转换、聚合等），然后是长期存储。可视化工具从存储单元中获取分析结果，并以图形、警报、图表、Excel 表、仪表板或地图的形式将其呈现给商业所有者，这些所有者可以分析信息并根据信息采取一些行动。

2.2 NRT 的高级系统视图

在本章前面,我们特意给读者讲解了 NRT 应用程序的基本构建模块及其逻辑概述。下一步是理解 NRT 框架的功能和系统视图。图 2-6 清楚地概述了各种体系的构建模块和贯穿各领域的问题。

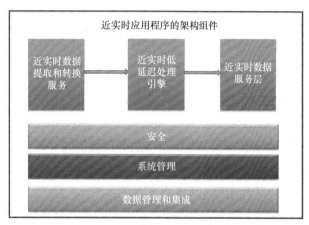

图 2-6

从图 2-6 中可以看到由左到右的横向流程,整个过程从使用低延迟组件数据提取和转换开始,它是近实时的,转换后的数据被传递到下一个逻辑单元,该逻辑单元实际上对数据执行了高度优化和并行操作——该单元实际上是近实时处理引擎。一旦数据被聚合和关联,并且得到了可行的分析结果,那么它就被传递到表现层。与实时仪表板和可视化工具组合在一起,表现层可能有一个持久性组件,以用于长期深入分析保留数据。

如图 2-6 所示,NRT 框架所有组成部分中贯穿各领域的问题包括安全、系统管理以及数据完整性和管理。

接下来,我们将带领读者了解 4 种基本的流处理模式,帮助读者了解流处理用例构成的常见风格及最佳解决方案(见本章后续内容)。

- **流提取**:在这里,我们所要做的就是将事件持久化到稳定的存储中,如 HDFS、HBase、Solr 等。所以需要的只是低延迟的流收集、转换和持久性组件。
- **近实时(NRT)处理**:此应用程序设计允许有外部上下文,并处理异常或欺诈检测等复杂用例。它需要根据特定的复杂业务逻辑对事件进行过滤、警报、重复数据消除和转换。所有这些操作都需要在极低的延迟下执行。

- **NRT 事件分区处理**：这与 NRT 处理非常接近，但是有一个变化可以帮助它从分区数据中获得好处，从而引用一些实例，这就像在内存中存储相关的外部信息。这种模式也应在极低的延迟下运行。

- **NRT 和复杂模型/机器学习**：它主要要求我们在流事件集的滑动时间窗口上执行非常复杂的模型/操作。它们是高度复杂的操作，需要对数据进行微批，并且要求操作延迟非常低。

2.3 NRT 的技术视图

在本节中，我们将向读者介绍 NRT 组件的各种技术选型及其在特定情形下的优缺点。随着本书的进展，之后我们将更详细地介绍这些内容，以帮助读者理解为什么某些工具和技术栈更适合解决某些用例。

在此之前，我们有必要了解一些关键的相关工具和技术。这里提到的工具和技术对于软件来说是通用的，我们稍后将继续讨论 NRT 工具的细节。

- **性能**：基本上它是在给定负载和给定硬件环境下测量软件组件的性能。
- **容量**：这一点非常关键，因为它决定了任何应用程序的断点。
- **管理**：组件的管理烦琐吗？是否需要专业工程师来维护？
- **可扩展性**：组件是否可扩展，以适应不断增加的流量和新的用例？能在同一个体系上演进吗？
- **总拥有成本**（TCO）：端到端解决方案所需的总成本。

既然有了经验法则，那么让我们看看 NRT 应用程序框架的技术视图，了解相关的技术选型，如图 2-7 所示。

图 2-7 展示了 NRT 框架在设计解决方案时的各种关键技术，下面我们进一步了解每个技术环节及其候选的技术选择。

2.3 NRT 的技术视图

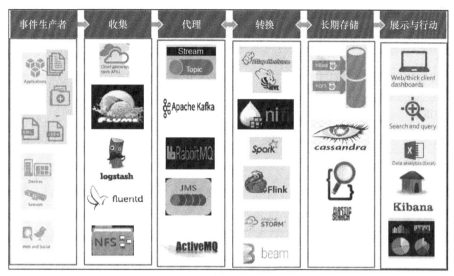

图 2-7

2.3.1 事件生产者

事件生产者是事件发生的源头。这些单独的事件或元组以连续永不结束的流串接，形成数据流，这些数据流作为任何 NRT 应用程序系统的输入源。事件可以是以下任意一个或多个事件，并且可以实时触发。

- 应用程序、电信日志或者 CDR。
- Facebook、Twitter 或者其他社交媒体。
- 传感器或者卫星产生的数据。
- 闭路电视或者安全摄像头的视频。
- 图像、自由文本、结构化、半结构化、非文本（音频、视频）、JSON、XML 等格式的数据。

2.3.2 数据收集

既然已经确定了数据的来源及其特征和到达频率，接下来考虑将实时数据导入应用程序中的各种收集工具。各种数据收集工具的比较见表 2-1。

表 2-1

	FluentD	Flume	Logstash
安装方式	Gem/rpm/deb	Jar/rpm/deb	apt/yum
代码量	3000 行 Ruby	50000 行 Java	5000 行 JRuby
插件开发语言	Ruby	Java	JRuby
插件管理	RubyGems.org	无	Logstash-plugins
是否有主服务器	否	是的	是的
许可证	Apache	Apache	Apache
是否具备可扩展性	是	是	是
是否为分布式	是	是	是

（1）**Apache Flume**：Flume 是一种分布式、高可靠和高可用的服务，用于高效地收集、聚合和移动大量日志数据。它有一个简单且灵活的基于数据流的体系结构，具有鲁棒性和容错性、可维护性、可靠性机制以及故障转移和恢复机制。它使用一个简单的可扩展数据模型，允许在线分析应用程序。Flume 的主要特点如下。

- **简单地读取流数据，具有内置故障恢复功能**。通过内存和磁盘信道处理激增或峰值数据，而不会影响下游处理系统。

- **消息传递有保障**：它有一个内置的信道机制，可以在确认上工作，从而确保传递消息。

- **可扩展性**：与所有其他 Hadoop 组件一样，Flume 很容易水平扩展。

（2）**FluentD**：FluentD 是一个开源数据采集器，统一数据收集和消费，以便更好地使用和理解数据。FluentD 的显著特点如下。

- **可靠性**：此组件同时具有基于内存和文件的信道配置，可以根据所需用例的可靠性需求进行配置。

- **基础架构采用底层语言编写**：组件用 Ruby 和 C 编写，内存和 CPU 消耗非常少。

- **可插拔架构**：该组件为社区的不断发展做出了贡献。

- **使用 JSON**：它将数据尽可能地统一为 JSON，从而使统一、转换和筛选变得更容易。

（3）**Logstash**：Logstash 是一个开源的服务器端数据处理管道，它同时从多个源接收、转换数据。它的显著特点如下。

- **多样性**：支持各种各样的流数据输入源，包括从度量到应用程序日志、实时传感器数据、社交媒体数据，等等。
- **过滤输入数据**：Logstash 具备超低延迟的分析、过滤和转换数据的能力。在某些情况下，我们可能希望在登录到代理或存储之前，对来自各种源的数据进行筛选，并按照预先定义的通用格式进行解析。这使得整体开发过程分离，并且由于它收敛到通用格式而易于使用。它能够格式化和解析非常复杂的数据，并且整个处理时间与源、格式、复杂性或模式无关。
- 可以将转换后的输出统一收集到各种存储、处理或下游应用系统（如 Spark、Storm、HDFS、ES 等）中。
- **良好的鲁棒性、功能强大且可扩展**：开发者可以选择使用各种各样的可用插件或编写自定义插件。插件可以使用名为**插件生成器**的 Logstash 工具开发。
- **监控 API**：开发人员能够利用 Logstash 集群并监控数据管道的整体运行状况。
- **安全性**：有提供动态加密数据的能力，以确保数据安全。

（4）**用于数据收集的云 API**：这是另一种数据收集方法，大多数云平台都提供各种数据收集 API，例如，亚马逊 Kinesis Firehose、Google StackDriver 监控 API、数据收集 API 和 IBM Bluemix 数据连接 API。

2.3.3 代理

组件分离是一个基本的架构原则。代理恰恰是 NRT 体系结构中的一个组件，它不仅将数据采集组件和处理单元分离开，还提供了在流量突然激增时将数据保留在队列中的弹性。

在本节中，我们介绍的各种工具和技术主要如下。

（1）**Apache Kafka**：Kafka 用于构建实时数据管道和流式应用程序。它具有水平可扩展性、容错性和非常快的速度，并已在数千家公司的生产中运行。这个代理组件的显著特性如下。

- 高可扩展。

- 故障安全。
- 提供容错性和高可用性。
- 开放源代码和可扩展。
- 它是基于磁盘的,可以存储大量数据(这是 Kafka 的一个 USP,实际上可以满足任何数量的数据);允许数据的复制和分区。

(2)**ActiveMQ**:Apache ActiveMQ 速度快,支持多种跨语言客户端和协议,具有易于使用的企业集成模式和许多高级功能,同时完全支持 JMS 1.1 和 J2EE 1.4。Apache ActiveMQ 根据 Apache 2.0 许可证发布。该协议的主要特点如下。

- 支持多种客户端和协议。
- 支持 JMS 1.1 和 J2EE 1.4。
- 高性能。
- 支持持久性。
- 公开了与技术无关的 Web 服务层。

(3)**RabbitMQ**:这是一个具有持久性、低延迟的位于内存中的分布式队列系统,具有以下显著功能。

- 鲁棒性。
- 易于使用。
- 支持所有主要操作系统。
- 既有开源版本,也有商业版本。

其中 RabbitMQ 和 Apache Kafka 的对比,即两种消息队列组件的对比见表 2-2。

表 2-2

	RabbitMQ	**Apache Kafka**
定义	通用、低延迟的代理系统,支持多种工业级的协议类型,如 AMQP	企业级服务总线,为高性能、低延迟数据流系统而创建和优化
许可证	Mozilla Public License	Apache License
开发语言	Erlang	Scala(JVM)

续表

	RabbitMQ	**Apache Kafka**
是否支持高可用性	支持	支持
是否支持联合队列	支持	不支持
是否支持复杂路由	支持	不支持
是否支持可扩展性	支持水平扩展	支持垂直扩展

2.3.4 转换和处理

转换和处理是 NRT 框架的核心,所有处理实际上都在这里执行,从数据转换到复杂的聚合和其他操作。

可供选择的组件有 Apache Storm、Apache Flink 和 Apache Spark。在接下来的章节中,我们将对它们进行详细的探讨。这些具备可扩展、分布式且高可用的 NRT 核心框架的特性对比见表 2-3。

表 2-3

	Apache Storm	**Apache Spark**	**Apache Flink**
是否支持流式处理	支持		支持
API		高级	高级
是否有容错性	有(tuple ack)	基于 RDD(谱系)	粗粒度检查点
是否有状态			内部有状态
是否支持精确一次语义		支持	支持
滑动窗口			灵活
低延迟	很低		低
高吞吐量	良好	高	高

2.3.5 存储

首先,我们需要将中间或最终结果和警报写入稳定的存储空间。存储是 NRT 应用程序中非常重要的组件,因为我们需要将最终结果进行持久化存储。其次,它是下游应用程序的集成点,这些应用程序从这些低延迟存储中提取数据,并对它们进行进一步的洞察或深入的学习。

图 2-8 清楚地表示了各种数据存储及其与 NRT 应用程序的时间 SLA 的一致性。

虽然现在跳过了大量可用于存储和可视化的选择方案,但我们将在本书的后面具体介绍这些选择方案。

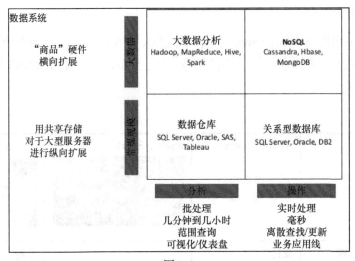

图 2-8

2.4 小结

在本章,我们介绍了 NRT 体系结构框架的各种组件和技术选型。读者了解了实时处理的挑战、要考虑的关键方面以及堆栈中每种技术可用的 USP。这里的目的是让读者从概念上熟悉可用工具和技术栈选择,以便根据功能性和非功能性需求,选择最适合用例的解决方案。

第二部分 搭建基础设施

- 第 3 章　了解和跟踪数据流
- 第 4 章　安装和配置 Storm
- 第 5 章　配置 Apache Spark 和 Flink

第 3 章
了解和跟踪数据流

在本章，我们将深入探讨实时应用程序的核心技术，即从源获取流数据到计算组件。我们将讨论需求及一些可用的技术选型，还将向读者说明哪些技术更适合某些用例和场景，从基本用例到高级设置均有涉及。本章还将介绍与用例的数据提取相关的技术。

── 本章主要包括以下内容 ──

- 理解数据流
- 为数据提取安装基础设施
- 将数据从源填到处理器——需求和注意事项
- 比较与选择适合用例的最佳实践
- 小试牛刀

3.1 了解数据流

数据流是使用任何介质的任何类型数据的连续流。在大数据的"4V"概念中，有两个"V"分别指的是速度和各种数据。数据流是来自社交媒体网站或安装在企业或车辆上的不同监控传感器等源中的实时数据。流式数据处理的另一个例子是物联网，数据通过互联网来自不同的组件。大数据处理中的实时数据流如图 3-1 所示。

有两种不同类型的流数据：有界流和无界流，如图 3-2 所示。有界流具有定义的开始和结束。一旦到达流的结尾，数据处理就停止。通常，这称为批处理。无界流没有结束，数据从开始就进行处理。这称为实时处理，它将事件的状态保存在内存中以进行处理。管理和实现无界数据流的用例是非常困难的，但是有一些工具可以让读者有机会使用它们，包括 **Apache Storm**、**Apache Flink**、**Amazon Kinesis** 和 **Samaza** 等。

在接下来的章节中，我们将讨论数据处理。这里主要讨论的数据提取工具会集成到数据处理引擎中。数据可能来自生成日志文件的系统，或者直接来自终端或端口。

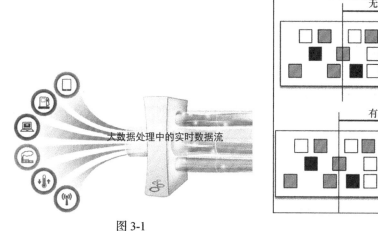

图 3-1

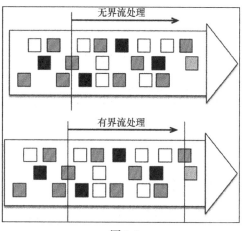

图 3-2

3.2 为数据提取安装基础设施

数据提取的工具和框架有许多，本节将主要集中讨论 Apache Kafka、Apache NiFi、Logstash、Fluentd 和 Flume 这几个工具。

3.2.1 Apache Kafka

Kafka 是一种消息代理，可以连接到现有的任何实时框架。本书中，Kafka 将经常出现在各种类型的用例中。在本节中，我们用 Kafka 作为数据源，将来自文件的数据转换为队列形式，以供进一步处理。下载 Kafka 到本地计算机。下载 `kafka_2.11-0.10.1.1.tgz` 文件后，使用以下命令解压文件：

```
cp kafka_2.11-0.10.1.1.tgz /home/ubuntu/demo/kafka
cd /home/ubuntu/demo/kafka
tar -xvf kafka_2.11-0.10.1.1.tgz
```

解压缩后的文件和文件夹，如图 3-3 所示。

图 3-3

 用户可以在 server.properties 文件中更改侦听器的属性。注意它是纯文本形式：
PLAINTEXT//localhost:9092。

要启动 Kafka，请使用以下命令：

```
/bin/zookeeper-server-start.sh config/zookeeper.properties
/bin/kafka-server-start.sh config/server.properties
```

Kafka 会在用户的本地机器上启动。稍后我们将根据需要创建 topic。接下来介绍 NiFi 的安装与应用。

3.2.2 Apache NiFi

Apache NiFi 是从源文件中读取数据并在不同类型的接收器之间分发数据的工具。有多种类型的源和接收器连接器可用。将 NiFi 1.1.1 版下载到本地计算机。下载 nifi-1.1.1-bin.tar.gz 文件后，解压缩以下文件：

```
cp nifi-1.1.1-bin.tar.gz /home/ubuntu/demo
cd /home/ubuntu/demo
tar -xvf nifi-1.1.1-bin.tar.gz
```

解压缩的文件目录，如图 3-4 所示。

图 3-4

用如下命令启动 NiFi：

```
/bin/nifi.sh start
```

NiFi 是在后台启动的。要检查 NiFi 是否运行成功，可使用以下命令：

```
/bin/nifi.sh status
```

启动 NiFi 后，读者可以通过访问 http://localhost:8080/nifi 来访问 NiFi UI。图 3-5 显示了 NiFi UI。

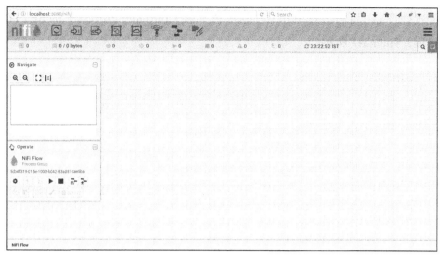

图 3-5

现在，让我们在 NiFi 中创建一个流文件，它将读取文件并将每一行作为名为 nifi-example 的 Kafka 主题中的一个事件进行推送。首先，使用以下命令在 Kafka 中创建主题：

```
/bin/kafka-topics.sh --create --topic nifi-example --zookeeper localhost:2181 --partitions 1 --replication-factor 1
```

 读者应该在/etc/hosts 中正确输入系统的 IP 地址。否则，在 Kafka 中创建主题时会遇到问题。

现在，读者将注意力转到 NiFi UI。选择 Processor 并将其拖到窗口中。在 NiFi 中将显示所有可用的处理器。搜索 GetFile 并选择它，它将显示在读者的工作区中，如图 3-6 所示。

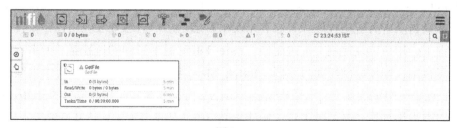

图 3-6

要配置处理器，可右击 GetFile 处理器，然后在 Configure Processor 的 SETTINGS 页面进行设置，如图 3-7 所示。

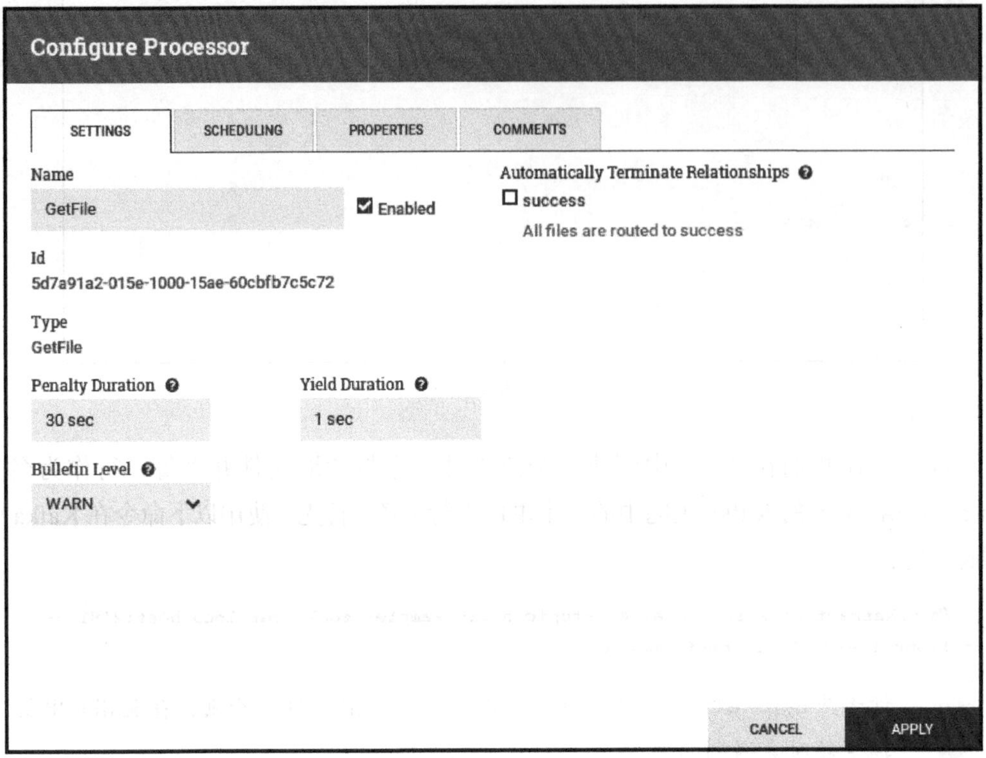

图 3-7

按照上面的配置方法，读者能够灵活地更改与处理器类型相关的所有可能配置。鉴于本章的范围，直接跳转到属性。应用图 3-8 所示的属性。

Input Directory 是保存日志/文件的目录。**File Filter** 是筛选出目录中文件的正则表达式。假设目录中有应用程序级日志和系统级日志，用户只想处理应用程序级日志。在这种情况下，用户可以使用文件过滤器。**Path Filter** 是子目录的过滤器。如果日志目录有多个子目录，则可以使用此过滤器。**Batch Size** 是正在运行的一次迭代中提取的最大文件数。如果不想删除源文件，请将 **Keep Source File** 设置为真。**Recurse Subdirectories** 是每当需要扫描日志目录中的子目录时使用的属性。如果需要，则将其设置为真；否则，将其设置为假。**Polling Interval** 是进程在日志目录中查找新文件的时间。如果要处理日志目录中的隐藏文件，请将 **Ignore Hidden Files** 设置为假。

3.2 为数据提取安装基础设施

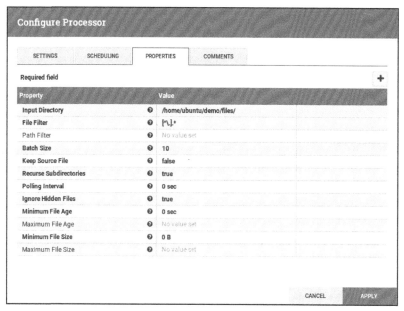

图 3-8

要读取我们使用的 **Getfile** 处理器的文件,现在要在 Kafka 主题推送每一行,然后使用 **PutKafka** 处理器。再次单击处理器,并将其拖到工作区域。

鼠标按下后,它将询问处理器的类型。搜索处理器作为 **PutKafka** 并选择它,它将显示在图 3-9 中。

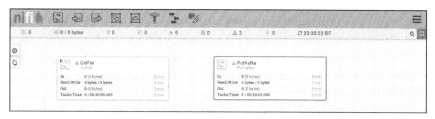

图 3-9

现在,右键单击 **PutKafka** 并选择配置进行配置。按图 3-10 所示的内容设置配置。

一些重要的配置是 **Known Brokers**、**Topic Name**、**Partition** 和 **Client Name**。

可以在已知代理中指定代理主机名和端口号,多个代理用逗号分隔。指定在 Kafka 代理上创建的主题名。分区是在对主题进行分区时使用的。**Client Name** 应该是客户机与 Kafka 建立连接的任何相关名称。

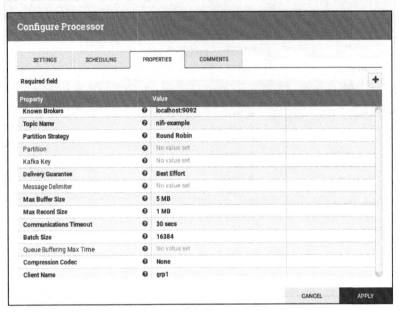

图 3-10

现在，在 **GetFile** 处理器和 **PutKafka** 处理器之间建立连接。将箭头从 **GetFile** 处理器拖动到 **PutKafka** 处理器，以创建连接。

在文件/home/ubuntu/demo/和一些单词或语句中创建测试文件，如下：

```
hello
this
is
nifi
kafka
integration
example
```

在运行 NiFi 管道之前，从控制台启动一个进程以读取名为 nifi-example 的 Kafka 主题。

让我们启动 NiFi 管道，它读取测试文件并将其放入 Kafka 中。转到 NiFi 工作区，按键（Shift+A），然后在操作窗口按**播放**按钮。输出如图 3-11 所示。

图 3-11

NiFi 输出如图 3-12 所示。

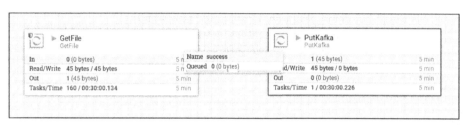

图 3-12

3.2.3 Logstash

下载 Logstash 的 5.2.1 版本到本地机器，并下载文件 **logstash-5.2.1.tar.gz**，然后通过如下命令提取文件：

```
cp logstash-5.2.1.tar.gz ~/demo/.
tar -xvf ~/demo/logstash-5.2.1.tar.gz
```

解压缩的文件目录如图 3-13 所示。

图 3-13

对 NiFi 也是采取相同的操作。读取一个文件，然后将事件上传到 Kafka 主题。为 Logstash 创建一个 Kafka 主题：

```
./kafka-topics.sh --create --topic logstash-example --zookeeper localhost:2181 --partitions
```

先创建一个配置文件，然后才能在 Logstash 中运行该示例。该配置文件定义了输入、过滤器和输出。此处不会应用任何过滤器，因此配置文件将包含两个组件：读取文件的输入和写入 Kafka 主题的输出。以下是执行示例所需的配置文件。

```
input {
    file {
        path => "/home/ubuntu/demo/files/test"
        start_position => "beginning"
    }
}
output {
    stdout { codec => plain }
    kafka {
        codec => plain {format => "%{message}"}
        topic_id => "logstash-example"
        client_id => "logstash-client"
        bootstrap_servers => "localhost:9092"
    }
}
```

创建名为 `logstash-kafka-example.conf` 的文件,并将之前的配置粘贴到该文件中。在 /home/ubuntu/demo/files 中创建一个名为 test 的输入文件,并添加以下内容:

```
hello
this
is
logstash
kafka
integration
```

在运行 Logstash 管道之前,使用以下命令从控制台启动一个进程来读取 Kafka 主题中 `logstash-example` 的内容。

```
/bin/kafka-console-consumer.sh --topic logstash-example --bootstrap-server localhost:9092
```

然后,使用如下命令运行实例。

```
/bin/logstash -f config/logstash-kafka-example.conf
```

Kafka 的输出如图 3-14 所示。

图 3-14

3.2.4 Fluentd

Fluentd 是另一种处理日志文件的工具。Fluentd 有 3 个组件，与 Logstash 中的组件相同，分别是输入、过滤器和输出。根据用户的用例需要，有多个输入和输出插件可用。此处，本书将演示一个与前面类似的例子，即读取日志文件并将其写入 Kafka。

下载 Fluentd。因为我们使用的是 Ubuntu，所以选择 Debian 安装。下载 `td-agent_2.3.4-0_amd64.deb`，并使用软件中心在 Ubuntu 中安装它。

它在系统上安装后，使用以下命令对其进行验证：

```
sudo td-agent --dry-run
```

执行上述命令，若生成如下信息，说明一切正常：

```
2017-02-25 16:19:49 +0530 [info]: reading config file path="/etc/tdagent/td-agent.conf"
2017-02-25 16:19:49 +0530 [info]: starting fluentd-0.12.31 as dry run mode
2017-02-25 16:19:49 +0530 [info]: gem 'fluent-mixin-configplaceholders'version '0.4.0'
2017-02-25 16:19:49 +0530 [info]: gem 'fluent-mixin-plaintextformatter'version '0.2.6'
2017-02-25 16:19:49 +0530 [info]: gem 'fluent-plugin-kafka' version'0.5.3'
2017-02-25 16:19:49 +0530 [info]: gem 'fluent-plugin-kafka' version'0.4.1'
2017-02-25 16:19:49 +0530 [info]: gem 'fluent-plugin-mongo' version'0.7.16'
2017-02-25 16:19:49 +0530 [info]: gem 'fluent-plugin-rewrite-tagfilter'version '1.5.5'
2017-02-25 16:19:49 +0530 [info]: gem 'fluent-plugin-s3' version'0.8.0'
2017-02-25 16:19:49 +0530 [info]: gem 'fluent-plugin-scribe' version'0.10.14'
2017-02-25 16:19:49 +0530 [info]: gem 'fluent-plugin-td' version'0.10.29'
2017-02-25 16:19:49 +0530 [info]: gem 'fluent-plugin-td-monitoring'version '0.2.2'
2017-02-25 16:19:49 +0530 [info]: gem 'fluent-plugin-Webhdfs' version'0.4.2'
2017-02-25 16:19:49 +0530 [info]: gem 'fluentd' version '0.12.31'
2017-02-25 16:19:49 +0530 [info]: adding match pattern="td.*.*"type="tdlog"
2017-02-25 16:19:49 +0530 [info]: adding match pattern="debug.**"type="stdout"
2017-02-25 16:19:49 +0530 [info]: adding source type="forward"
2017-02-25 16:19:49 +0530 [info]: adding source type="http"
2017-02-25 16:19:49 +0530 [info]: adding source type="debug_agent"
2017-02-25 16:19:49 +0530 [info]: using configuration file: <ROOT>
  <match td.*.*>
    @type tdlog
    apikey xxxxxx
    auto_create_table
    buffer_type file
    buffer_path /var/log/td-agent/buffer/td
    buffer_chunk_limit 33554432
    <secondary>
      @type file
```

```
      path /var/log/td-agent/failed_records
      buffer_path /var/log/td-agent/failed_records.*
    </secondary>
  </match>
  <match debug.**>
    @type stdout
  </match>
  <source>
    @type forward
  </source>
  <source>
    @type http
    port 8888
  </source>
  <source>
    @type debug_agent
    bind 127.0.0.1
    port 24230
  </source>
</ROOT>
```

在 Fluentd 中要创建管道，需要编写 Fluentd 可读的配置文件，并处理管道。Fluentd 配置文件的默认位置是 /etc/td-agent/td-agent.conf。下面是从日志文件中读取并将每个事件推送到 Kafka 主题中的配置文件：

```
<source>
  @type tail
  path /home/ubuntu/demo/files/test
  pos_file /home/ubuntu/demo/fluentd/test.log.pos
  tag fluentd.example
  format none
</source>
<match *.**>
  @type kafka_buffered
  brokers localhost:9092
  default_topic fluentd-example
  max_send_retries 1
</match>
```

在配置文件中，前面引用了 6 个可用指令中的 2 个。Source 指令是所有数据的来源。@type 告诉用户使用的是哪种类型的输入插件。这里使用 tail，它将跟踪日志文件。这对于输入日志文件正在运行日志文件的用例很合适，其中事件/日志是在文件末尾追加的。它与 Linux 中的 tail -f 操作相同。尾部输入插件有多个参数。Path 是日志文件的绝对路径。`Pos_file` 是记录输入文件最后读取位置的文件。`Tag` 是事件的标记。如果要定义输入格式（如 CSV）或应用正则表达式，请使用 format 命令。由于 format 参数是必需的，因此其参数取为 None，即使用该值将按原样使用输入文本。

`Match` 指令告诉 Fluentd 如何处理输入，*.The**模式告诉我们，无论日志文件中有什么内容，只要将其推送到 Kafka 主题即可。如果要对错误和信息日志使用不同的主题，请将模式定义为错误或信息，并将输入标记为相同。broker 参数表示 Kafka Broker 在系统上运行的主机和端口。默认主题是要在其中推送事件的主题。如果需要在消息失败后重试，请将 `max_send_retries` 设置为一个或多个。

将/etc/td-agent/td-agent.conf 替换为之前的配置，创建 Kafka 主题。

```
/bin/kafka-topics.sh --create --topic fluentd-example --zookeeper localhost:2181 --partitions
```

启动 Fluentd 代理：

```
sudo td-agent
```

启动 Kafka 消费者：

```
/bin/kafka-console-consumer.sh --topic fluentd-example --bootstrap-server localhost:9092
```

若启动该过程没有异常，请在/home/ubuntu/demo/files/test 中添加语句，如图 3-15 所示。

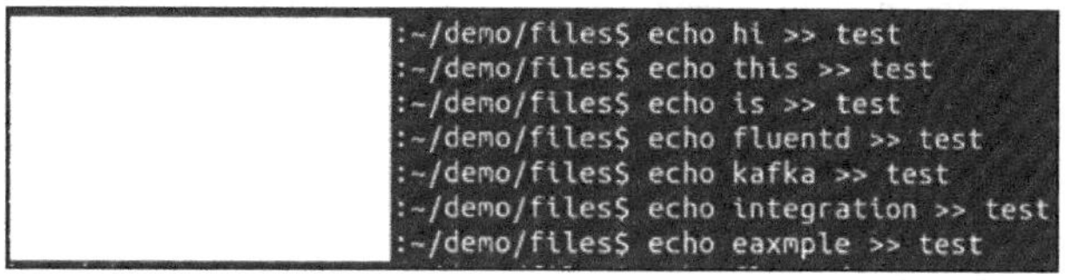

图 3-15

Kafka 输出如图 3-16 所示。

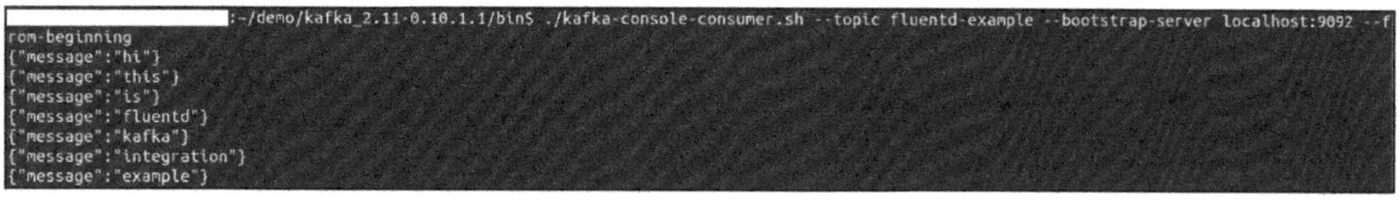

图 3-16

3.2.5 Flume

Flume 是 Apache 日志处理开源项目中最著名的一个。下载 `apache-flume-1.7.0-bin.tar.gz` 安装文件并解压缩。

```
cp apache-flume-1.7.0-bin.tar.gz ~/demo/
tar -xvf ~/demo/apache-flume-1.7.0-bin.tar.gz
```

解压缩后的文件目录如图 3-17 所示。

```
:~/demo/apache-flume-1.7.0-bin$ ls
bin  CHANGELOG  conf  DEVNOTES  doap_Flume.rdf  docs  lib  LICENSE  logs  NOTICE  README.md  RELEASE-NOTES  tools
```

图 3-17

如同前面的工具一样，接下来将演示类似的执行示例，包括读取文件和推送到 Kafka 主题。首先配置 Flume 文件：

```
a1.sources = r1
a1.sinks = k1
a1.channels = c1
a1.sources.r1.type = TAILDIR
a1.sources.r1.positionFile = /home/ubuntu/demo/flume/tail_dir.json
a1.sources.r1.filegroups = f1
a1.sources.r1.filegroups.f1 = /home/ubuntu/demo/files/test
a1.sinks.k1.type = org.apache.flume.sink.kafka.KafkaSink
a1.sinks.k1.kafka.topic = flume-example
a1.sinks.k1.kafka.bootstrap.servers = localhost:9092
a1.channels.c1.type = memory
a1.channels.c1.capacity = 1000
a1.channels.c1.transactionCapacity = 6
a1.sources.r1.channels = c1
a1.sinks.k1.channel = c1
```

Flume 通过 3 个组件来定义流。第一个组件是源，也就是日志/事件的来源。Flume 允许用多个源定义流，除了 Kafka、TAILDIR 和 HTTP，用户还可以定义自己的源。第二个组件是接收器，也就是事件最终被消费的目的地。第三个组件是信道，它定义了源和接收器之间的介质。最常用的信道是内存、文件和 Kafka，但也有更多其他的信道。在这里，用户将使用 TAILDIR 作为源，Kafka 作为接收器，而内存作为信道。与以前一样，配置 a1 为代理名称，r1 为源，k1 为接收器，c1 为信道。

从源的配置开始。首先，必须使用<agent-name>.<sources/sinks/channels>.<alias name>.type 来定义源的类型。用参数 positionFile 来跟踪日志文件。filegroups 指示要跟踪的一组文件。filegroups.<filegroup-name> 是文件夹的绝对路径。接收器的配置简单直接。Kafka 需要引导服务器和主题名称。信道的配置很长，但在这里我们只使用其中最重要的一些。容量是在信道中存储最多的事件数，事务容量是每个事务通道将从源处获取或提供给接收器的最多事件数。

接下来，通过如下命令开启 Flume agent。

```
bin/flume-ng agent --conf conf --conf-file conf/flume-conf.properties --name a1 -Dflume
```

Flume agent 将启动并输出图 3-18 所示的内容。

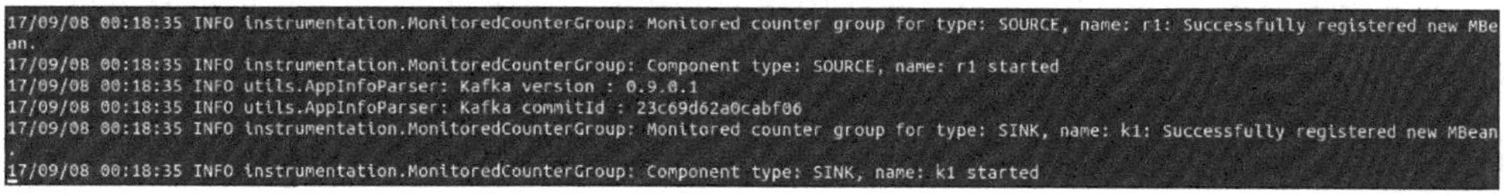

图 3-18

创建 Kafka 主题并命名为 `flume-example`：

```
bin/kafka-topics.sh --create --topic flume-example --zookeeper localhost:2181 -partitions
```

然后，启动 Kafka 控制台消费者：

```
bin/kafka-console-consumer.sh --topic flume-example --bootstrap-server localhost:9092
```

然后，将推送一些消息到文件 /home/ubuntu/demo/files/test 中，如图 3-19 所示。

图 3-19

Kafka 的输出如图 3-20 所示。

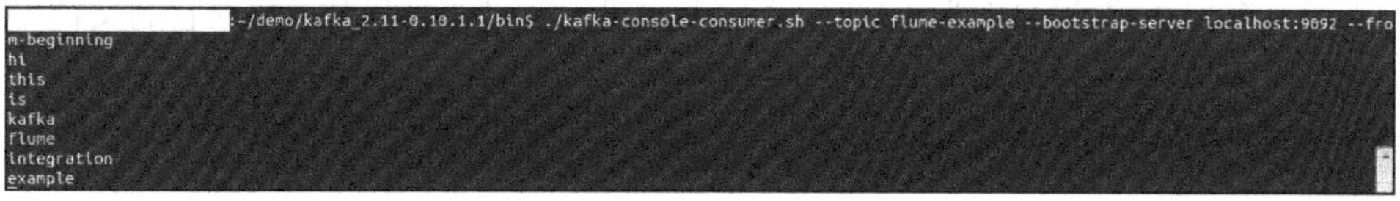

图 3-20

3.3 将数据从源填到处理器——期望和注意事项

在本节，我们将讨论日志流工具在性能、可靠性和可扩展性方面的特点。系统的可靠性可以通过消息传递语义来确定。传递语义包括如下 3 种类型。

- **最多一次**：消息立即传输。如果传输成功，消息将不再发送出去。但是，许多失败情况都会导致消息丢失。
- **至少一次**：每条消息至少传输一次。在失败的情况下，消息可能会被传输两次。
- **只有一次**：每条消息只传输一次。

性能包括 I/O、CPU 和 RAM 的使用情况和影响。根据定义，可扩展性是系统、网络或进程处理不断增加的工作量的能力，或者为了适应这种增长而需要扩展的潜力。因此，本书接下来将确定前文提到的一些工具是否具备可扩展性以处理增加的负载。可扩展性可以在水平和垂直方向上实现。水平扩展意味着增加更多的计算机器并分配工作，而垂直扩展意味着一台机器在 CPU、RAM 或 IOPS 方面增加的容量。

本节从 NiFi 开始说起。默认情况下，它是一个有保证的传输处理引擎（只有一次），它维护预写日志和实现此目的的内容存储库。性能取决于我们选择的可靠性。在 NiFi 保证消息传递的情况下，所有消息都将写入磁盘，然后从磁盘中读取。这将拖慢速度，但如果你不想失去任何一条信息，必须在性能方面付出代价。我们可以创建一个由 NiFi 集群管理器控制的 NiFi 集群。在内部，它由 Zookeeper 管理，以同步所有节点。模型是主节点和从节点，但是如果主节点死了，那么所有节点都会继续运行。一个限制是没有新的节点可以再加入集群，并且读者不能更改 NiFi 流。因此，NiFi 具有足够的可扩展性来处理集群。

Fluentd 提供了**最多一次**和**至少一次**传输语义。可靠性和性能是通过使用缓冲区插件来实现的。内存缓冲区结构包含一个块队列。当顶部块超过指定的大小或时间限制时，新的空块将被推送到队列顶部。当一个新块被推入时，底部的块立即被写出。文件缓冲区提供了一个持久缓冲区实现，它使用文件在磁盘上存储缓冲区块。根据它的文档可知，Fluentd 是一个扩展性很好的产品，其中 $M*N$ 由 $M+N$ 解析，其中 M 是输入插件的数量，N 是输出插件的数量。通过配置多个日志转发器和日志聚合器，我们可以实现可扩展性。

Logstash 的可靠性和性能是通过使用第三方消息代理实现的。Logstash 最适合 Redis。

其他输入和输出插件可用于与消息代理集成,如 Kafka、**RabbitMQ** 等。以 `Filebeat` 作为叶节点,我们可以用 Logstash 获得可扩展性,以便从多个源、具有不同日志目录的同一源或具有不同文件过滤器的相同源和相同目录上进行读取。

Flume 使用信道和接收器与信道之间的同步来获得可靠性。接收器仅在事件存储到下一个代理信道或存储在终端存储库之后,才从信道中删除事件。这是单跳消息传递语义。Flume 使用事务性方法来保证事件的可靠传递。要通过网络读取数据,Flume 与 `Avro` 和 `Thrift` 集成。

3.4 比较与选择适合用例的最佳实践

Logstash、Fluentd、Apache Flume、Apache NiFi 和 Apache Kafka 之间的比较见表 3-1。

表 3-1

	Logstash	Fluentd	Apache Flume	Apache NiFi	Apache Kafka
缺点	没有 UI,过滤器难写	Windows 版本仍在实验阶段	难以管理多个连接	相比于市面上其他工具,没有那么成熟	没有 UI,难以维护偏移量
优点	灵活和可以互操作	简单、鲁棒性好	提供与 HDFS 最好的集成,可靠、易扩展	流程管理,易于使用,安全,灵活扩展模型	快速,基于数据流的发布/订阅,易于集成和使用,分区
插件	90 多个插件	125 多个插件	50 多个插件,支持自定义的组件	有足够的处理器可供使用,也可以自定义	没有插件,可以写代码来扩展
可扩展性	支持	支持	支持	支持	支持
传递语义	至少使用一次 Filebeat	至多一次、至少一次	至少一次(支持事务)	保证一次(默认)	通过配置支持至少一次、至多一次和保证一次

3.5 小试牛刀

在本节,我们将提出一些问题,以便读者可以在读完前面的内容之后创建自己的应用程序。

我们还将扩展前面给出的有关设置和配置 NiFi 的示例,从实时日志文件中读取问题并将其放入 Cassandra 中。伪代码为如下所示几个部分:

Tail 日志文件、将事件放入 Kafka 主题、阅读 Kafka 主题中的事件、过滤事件、将事件推送到 Cassandra。

首先安装 Cassandra 并对其进行配置,以便 NiFi 能够连接它。

Logstash 用于处理日志,并将它们扔到其他工具中进行存储或可视化。这里的最适合是**弹性搜索**、**Logstash** 和 **Kibana**(ELK)。本节建立弹性搜索和 Logstash 之间的集成,在下一章将集成弹性搜索和 Kibana,以实现完整的工作流程。构建 ELK 的所有操作如下。

- 创建从 PubNub 中读取实时传感器数据的程序。同一程序将把事件发布到 Kafka 主题。
- 在本地计算机上安装 Elasticsearch 并启动。
- 现在,编写一个从 Kafka 主题中读取内容的 Logstash 配置,解析和格式化它们,并将它们推入 Elasticsearch 引擎中。

设置 Elasticsearch

执行以下步骤以设置 Elasticsearch。

使用以下命令下载 Elasticsearch 安装包:

```
wget https://artifacts.elastic.co/downloads/elasticsearch/elasticsearch-5.2.2.tar.gz
```

使用以下命令提取 `elasticsearch-5.2.2.tar.gz`:

```
tar -xzf elasticsearch-5.2.2.tar.gz
```

进入 `elasticsearch-5.2.2 directory`:

```
cd Elasticsearch-5.2.2/
```

使用给定命令启动 Elasticsearch。

```
./bin/elasticSearch
```

3.6 小结

在本章，我们解释了数据流的概念，给出了相关的示例以及与数据流相关的实时用例，并向读者介绍了不同实时数据提取工具（如 Flume、NiFi、Logstash 和 Fluentd）的配置和快速执行。我们还解释了这些数据提取工具在可靠性和可扩展性方面所处的位置。然后，尝试比较数据提取工具，以便读者可以在比较利弊之后，根据用例的需要来选择这些工具。读者可以通过运行打包在 JAR 中的代码来运行示例——这些代码很容易在独立模式和集群模式下运行。最后，我们给读者提供了一个在使用伪代码的同时使用数据提取工具来解决的实时问题，这样就可以集中精力编写代码而不是找到正确的解决方案。

目前，读者已了解了不同类型的数据流工具。在第 4 章，我们将重点介绍 Storm 的设置。Storm 是一个开源的、分布式的、弹性的实时处理引擎，其设置包括下载、安装、配置和运行示例，以测试安装程序是否正常工作。

第 4 章 安装和配置 Storm

在本章，我们将引导读者安装和配置 Storm（包括单机模式和分布式模式），还会帮助读者写出并执行第一个 Storm 上的实时处理任务。

本章主要包括以下内容

- Storm 概述
- Storm 架构和组件
- 安装和配置 Storm
- 在 Storm 上实时处理任务

4.1 Storm 概述

Storm 是一款开源的、分布式的、弹性的实时处理引擎，由 Nathan Marz 于 2010 年末发明。Nathan Marz 当时在 **BackType** 工作，在博客中他提到当时构建 Storm 遇到的挑战。

最初，实时处理被实现为消息队列的形式，然后使用 Python 或任何其他语言从队列中读取消息并逐个处理它们。然而这一方法面临着如下几个挑战。

- 在任何消息处理失败的情况下，必须将其放回队列中以进行重新处理。
- 始终保持队列和处理单元正常运行。

以下是 Nathan 提出的两个令人振奋的想法，它们使 Storm 能够成为一个高度可靠的实时处理引擎。

- **抽象**：Storm 是流式的分布式抽象。流可以并行地生成和处理。spout 可以产生新的流，而 bolt 是流中的一个小处理单元。拓扑是顶级抽象。这里抽象的优点是不必担心内部正在发生的事情，例如，序列化/反序列化、在不同处理单元之间发送/接收消息等。用户可以专注于编写业务逻辑。
- 保证消息处理算法是第二个想法。Nathan 开发了一种基于随机数和 XOR 的算法，该算法只需要大约 20 字节就可以跟踪每个 spout 元组，而不管下游触发了多少个处理单元。

2011 年，BackType 被 Twitter 收购。Storm 在公开论坛上流行之后，开始被称为"实时 Hadoop"。2011 年 9 月，Nathan 正式发布了 Storm。2013 年 9 月，他正式提议在 Apache 孵化器中使用 Storm。2014 年 9 月，Storm 成为 Apache 的顶级项目。

4.2　Storm 架构和组件

在本节中，我们主要讨论 Storm 框架以及它的工作原理。Storm 集群如图 4-1 所示。

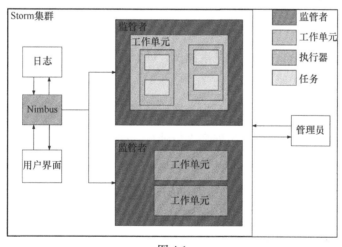

图 4-1

- **Nimbus** 节点充当 Storm 集群中的主节点。它负责分析拓扑结构，并根据任务的可用性将任务分配给不同的监管者。此外，它还负责监测故障。如果一个监管者死亡，它会将任务重新分配给可用的监管者。**Nimbus** 节点使用管理员（ZooKeeper）跟踪任务以保持其状态。如果 **Nimbus** 节点失败，它可以重新启动，以便读取管理员的状态，并从先前失败的点开始。

- 在 Storm 集群中，**监管者**（supervisor）充当从节点。一个或多个工作单元（即 JVM 进程），可以在每个监管者节点中运行。监管者与工作单元协调完成由 Nimbus 节点分配的任务。如果工作单元进程失败，监管者会找到可用的工作单元来完成任务。
- 工作单元(worker)是在监管者节点中运行的 JVM 进程，它拥有执行器(executor)。工作单元进程中可以有一个或多个执行器。工作单元与执行器协调完成任务。
- 执行器是由工作单元生成的单线程进程。每个执行器负责运行一个或多个任务。
- 任务（task）是单独的工作单元。它对数据执行实际处理。它可以是 spout，也可以是 bolt。
- 除了前面的进程，Storm 集群中还有两个重要组件：日志模块和用户界面。`logviewer` 服务用于调试 StormUI 上工作单元和监管者的日志。

4.2.1 特征

以下是 Storm 的重要特征。

- **快速**：根据 Hortonworks 的基准测试，每个节点每秒处理 100 万个 100 字节的消息，它轻量、快速。
- **可靠**：Storm 保证消息处理至少一次或最多一次。
- **可扩展**：Storm 可以每秒扩展到大量消息。这只需要添加更多个监管者节点，同时增加拓扑中的 spout 和 bolt 的并行度。
- **容错**：如果任何**监管者**死亡，则 Nimbus 节点将任务重新分配给另一个**监管者**。如果任何工作单元死亡，那么**监管者**将任务重新分配给另一个工作单元。这同样适用于**执行器**和**任务**。
- **与编程语言无关**：拓扑可以用任何语言编写。

4.2.2 组件

以下是 Storm 的组件。

- **元组**：这是 Storm 的基本数据结构。它可以包含多个值，每个值的数据类型可以不同。默认情况下，Storm 序列化原始类型的值，但如果读者有任何自定义类，则必须提供序列化程序并在 Storm 中注册它。元组提供了非常有用的方法，如 `getInteger`、`getString` 和 `getLong`，这样用户就不需要在元组中强制转换值了。

- **拓扑**：如前所述，拓扑是最高级别的抽象。它包含处理流程，流程中又包括 spout 和 bolt。这是一种图计算，每个流程以图的形式表示。因此，节点是 spout 或 bolt，边是连接它们的流分组。图 4-2 显示了一个简单的拓扑示例。

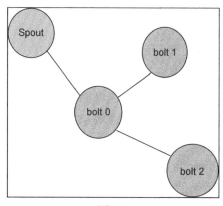

图 4-2

- **流**：流是 Storm 的核心抽象，它是一系列无界元组。流可以由不同类型的 bolt 来处理，从而产生新的流。每个流都提供一个 ID，如果用户未提供 ID，则 `default` 值是流的默认 ID。用户可以在 `OutputFieldDeclare` 类中定义流的 ID。

- **spout**：spout 是流的来源。它从诸如 Kafka、RabbitMQ 等源处读取消息作为元组并在流中把它们发出。spout 可以通过定义 `OutputFieldDeclare` 中的 `declareStream` 方法生成多个流。有如下两种类型的 spout。

    ◁ **可靠**：spout 跟踪每个元组并在发生任何故障时重放元组。

    ◁ **不可靠**：一旦将元组作为流发送到另一个 bolt 或 spout，spout 就不再关心元组。

以下是 Spout 的方法。

- **Ack**：在拓扑中成功处理元组时调用此方法。用户应将元组标记为已处理或已完成。
- **Fail**：未成功处理元组时调用此方法。用户必须以这样的方式实现此方法，即应该在 nextTuple 中再次发送元组以进行处理。
- **nextTuple**：调用此方法以从输入源处获取元组。从输入源读取的逻辑应该用这种方法写入并发送给元组以进一步处理。
- **Open**：初始化 spout 时，此方法仅调用一次。这里，与输入源或输出接收器建立连接或配置的内存高速缓存将确保它不会在 nextTuple 方法中重复。

IRichSpout 是 Storm 中可用于实现自定义 spout 的接口。上述所有方法都需要用户实现。

- **Bolt**：bolt 是 Storm 的处理单元。所有类型的处理（如过滤和聚合）都会加入数据库操作中。bolt 是一种转换，它将元组流作为输入并且不生成任何或更多的流作为输出。元组中的值类型或更多值可能也会发生变化。一个 bolt 可以通过定义 OutputFieldDeclare 的 declareStream 方法发出多个流。用户无法一次订阅所有流，必须逐个订阅它们。以下是 bolt 的方法。一个方法是 **Execute**。此方法在流中的每个元组上被执行其中 bolt 订阅它作为输入。在此方法中，可以通过转换数据库中的值或持久值来定义任何处理。对于处理的每个元组，一个 bolt 必须在 OutputCollector 上调用 ack 方法，以便 Storm 知道元组何时完成。另一个方法是 **Prepare**。当初始化 bolt 时，此方法只执行一次，因此无论什么连接或初始化类变量都可以进入此方法。

Storm 中提供了 IRichBolt 和 IBasicBolt 来实现 Storm 的处理单元。两者之间的区别在于 IBasicBolt 会自动识别每个元组，并提供基本的过滤器和简单的功能。

4.2.3 流分组

图 4-3 显示了 Storm 提供的不同类型的分组。

- **随机分组**：在任务期间随机分组，平均分配元组。所有任务都会收到相同数量的元组。
- **字段分组**：在此分组中，元组基于一个或多个字段被发送到同一个 bolt，例如，如果我们想要将来自同一条推文中的所有推文都发送到同一个 bolt，那么可以使用此分组。

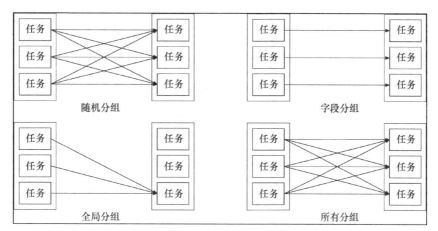

图 4-3

- **所有分组**：所有元组都被发送到所有 bolt。过滤是所有分组需要的一个操作。
- **全局分组**：所有元组都发送到一个 bolt。归纳是全局分组需要的一项操作。
- **直接分组**：元组的生产者决定消费者的哪个任务将接收元组。这仅适用于声明为直接流的流。
- **本地或随机分组**：如果源 bolt 和目标 bolt 在同一工作进程中运行，则它是本地分组，因为不需要通过网络路由跳来跨网络发送数据。如果不是这种情况，则与随机分组相同。
- **自定义分组**：用户可以定义自己的分组。

4.3 安装和配置 Storm

在安装 Storm 之前，首先需要先安装和设置 ZooKeeper。

4.3.1 安装 ZooKeeper

以下内容是有关如何在单机和集群模式下安装、配置和运行 ZooKeeper 的说明。

1. 安装

下载 ZooKeeper，然后解压缩 zookeeper-3.4.6.tar.gz，如下所示：

```
tar -xvf zookeeper-3.4.6.tar.gz
```

解压缩后的文件和文件夹如图 4-4 所示。

图 4-4

2. 配置

ZooKeeper 有两种类型的部署：单机部署和集群部署。两者的配置没有太大区别，只是集群模式有一些额外参数。

单机模式。如图 4-4 所示，进入 conf 文件夹并更改 zoo.cfg 文件，代码如下：

```
tickTime=2000 # Length of single tick in milliseconds. It is used to
# regulate heartbeat and timeouts.
initLimit=5 # Amount of time to allow followers to connect and sync
# with leader.
syncLimit=2 # Amount of time to allow followers to sync with
# Zookeeper
dataDir=/tmp/zookeeper/tmp # Directory where Zookeeper keeps
# transaction logs
clientPort=2182 # Listening port for client to connect.
maxClientCnxns=30 # Maximum limit of client to connect to Zookeeper
# node.
```

集群模式。除了上面的配置，集群模式下还要添加以下配置：

```
server.1=zkp-1:2888:3888
server.2=zkp-2:2888:3888
server.3=zkp-3:2888:3888
```

server.x = [hostname] nnnn: mmmm：这里 x 是分配给每个 ZooKeeper 节点的 id。在之前配置的 datadir 中，创建一个名为 myid 的文件，并在其中放入相应的 ZooKeeper ID。它在整个集群中应该是唯一的。这里使用相同的 ID 作为 x。nnnn 是跟随者用来连接领导节点的端口，mmmm 是用于选举领导者的端口。

3. 运行

使用以下命令从 ZooKeeper 主目录运行 ZooKeeper：

```
/bin/zkServer.sh start
```

控制台将在以下消息和过程在后台运行后出现。

```
Starting zookeeper ... STARTED
```

以下命令可用于检查 ZooKeeper 进程的状态：

```
/bin/zkServer.sh status
```

以下输出将处于单机模式：

```
Mode: standalone
```

以下输出将处于集群模式：

```
Mode: follower  # in case of follower node
Mode: leader    # in case of leader node
```

4.3.2 配置 Apache Storm

以下内容是有关如何使用 Nimbus 和监管者进行安装、配置和运行 Storm 的说明。

1. 安装

下载 Storm，然后解压缩文件 `apache-storm-1.0.3.tar.gz`：

```
tar -xvf apache-storm-1.0.3.tar.gz
```

解压缩后的文件和文件夹如图 4-5 所示。

图 4-5

2. 配置

如图 4-5 所示，进入 `conf` 文件夹并在 `storm.yaml` 中添加/编辑以下属性：

- 在 Storm 配置中设置 ZooKeeper 主机名。

```
storm.zookeeper.servers:
- "zkp-1"
```

```
    - "zkp-2"
    - "zkp-3"
```

- 设置 ZooKeeper 端口。

```
storm.zookeeper.port:2182
```

- 设置 Nimbus 节点主机名，以便 Storm 监管者可以与之通信。

```
nimbus.host: "nimbus"
```

- 设置 Storm 本地数据目录以保留 conf、JAR 等信息。

```
storm.local.dir: "/usr/local/storm/tmp"
```

- 设置将在当前监管者引擎节点上运行的工作单元数。最佳做法是使用与机器中核数相同的工作单元数量。

```
supervisor.slots.ports:
    - 6700
    - 6701
    - 6702
    - 6703
    - 6704
    - 6705
```

- 对工作单元、监管者和 Nimbus 执行内存分配。

```
worker.childopts: "-Xmx1024m"
nimbus.childopts: "-XX:+UseConcMarkSweepGC -
XX:+UseCMSInitiatingOccupancyOnly -
XX:CMSInitiatingOccupancyFraction=70"
supervisor.childopts: "-Xmx1024m"
```

- **拓扑的相关配置**：首先为要确认的元组树（完全处理过的）配置最长时间（以秒为单位）——在它被认为失败之前；然后将调试日志配置为 false，因此 Storm 将仅生成信息日志。

```
topology.message.timeout.secs: 60
topology.debug: false
```

3. 运行

启动完整的 Storm 集群需要如下 4 种服务。

- nimbus。首先，我们需要在 Storm 中启动 Nimbus 服务。以下是启动它的命令：

```
/bin/storm nimbus
```

- `supervisor`。接下来，我们需要启动监管者节点以连接 Nimbus 节点。以下是命令：

```
/bin/storm supervisor
```

- `ui`。要启动 Storm UI，请执行以下命令：

```
/bin/storm ui
```

用户可以访问 UI，如图 4-6 所示。

- `logviewer`。logviewer 服务有助于查看工作人员登录 Storm UI 的情况。执行以下命令启动它：

```
/bin/storm logviewer
```

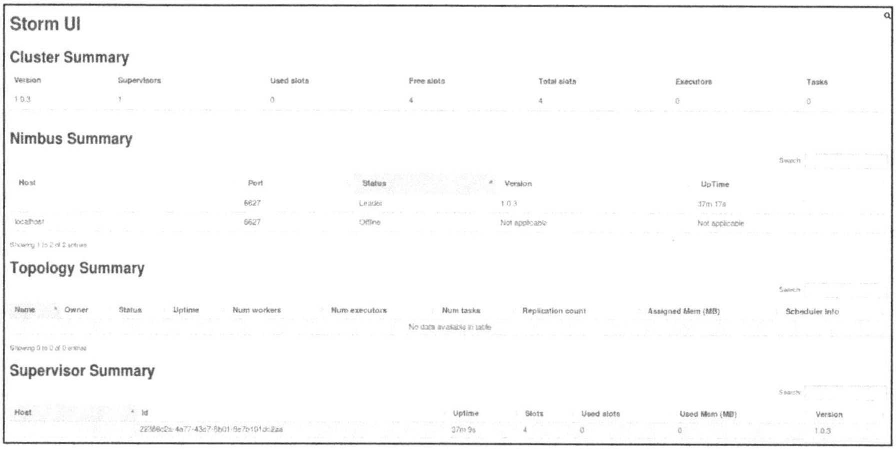

图 4-6

4.4 在 Storm 上实时处理任务

在讨论了 Storm 安装和配置的基础上，我们来看一个实时处理任务的示例。在本节，我们将讨论 Storm 中非常基本的例子——单词计数。为了在 Storm 中实现单词计数，我们们需要一个定期发出句子的 spout，一个 bolt 基于空格将句子分成单个单词，一个 bolt 收集所有单词并找到计数，最后需要一个 bolt 在控制台上显示输出。

下面来逐一讨论它们。

- **句子 spout**：要创建自定义 spout，首先必须扩展 `BaseRichSpout` 类，读者可以在其中提供所需方法的实现。创建固定的 spout，意味着每次迭代它都会发出相同的句子集，请创建句子的常量字符串数组。`declareOutputFields` 是定义流 ID 的方法，此流是 bolt 的输入。`nextTuple` 是迭代句子数组并将每个句子发送到下一个 bolt 的方法。

```java
public class FixedSentenceSpout extends BaseRichSpout {
    private static final long serialVersionUID = 1L;
    private SpoutOutputCollector collector;

private String[] sentences = { "This is example of chapter 4",
"This is word count example", "Very basic example of Apache
Storm","Apache Storm is open source real-time processing
engine"};
    private int index = 0;

    public void declareOutputFields(OutputFieldsDeclarer declarer) {
        declarer.declare(new Fields("sentence"));
    }

public void open(@SuppressWarnings("rawtypes") Map config,
TopologyContext context, SpoutOutputCollector collector) {
        this.collector = collector;
    }

        public void nextTuple() {
        String sentence = sentences[index];
        System.out.println(sentence);
        this.collector.emit(new Values(sentence));
        index++;
        if (index >= sentences.length) {
            index = 0;
        }
    }
}
```

- **拆分器 bolt**：首先，使用 `BaseRichBolt` 类扩展并实现所需的方法。在 `execute` 方法中，我们读取了一个带有 ID `Sentence` 的元组，它是我们在 spout 中定义的。然后，根据空格分割每个句子，并将每个单词作为元组发送到下一个 bolt。`declareOutputFields` 是一个用于定义每个 bolt 的流 ID 的方法。

```java
public class SplitSentenceBolt extends BaseRichBolt {
    private static final long serialVersionUID = 1L;
    private OutputCollector collector;

public void prepare(@SuppressWarnings("rawtypes") Map config,
TopologyContext context, OutputCollector collector) {
        this.collector = collector;
    }

    public void execute(Tuple tuple) {
        String sentence =
tuple.getStringByField("sentence");
        String[] words = sentence.split(" ");
        for (String word : words) {
            this.collector.emit(new Values(word));
        }
    }

    public void declareOutputFields(OutputFieldsDeclarer
declarer) {
        declarer.declare(new Fields("word"));
    }
}
```

- **字数计数 bolt**：在这个 bolt 中，我们维护一个映射，其中键为单词，值为计数。使用每个元组的一个 `execute` 方法计算该值。在 `declareOutputFields` 方法中，我们创建一个包含两个值的元组并将其发送到下一个 bolt。在 Storm 中，用户可以向下一个 bolt/spout 发送多个值，如以下示例所示。

```java
public void execute(Tuple tuple) {
    String word = tuple.getStringByField("word");
    Long count = this.counts.get(word);
    if (count == null) {
        count = 0L;
    }
    count++;
    this.counts.put(word, count);
    this.collector.emit(new Values(word, count));
}
public void declareOutputFields(OutputFieldsDeclarer declarer)
{
    declarer.declare(new Fields("word", "count"));
}
```

- **显示 bolt**：此 bolt 是拓扑中最后一个 bolt，因此，在 `declareOutputFields` 方法中无须定义任何内容。此外，`execute` 方法中不会发出任何内容。在这里，我们收集所有元组并将它们放入映射中。在拓扑终止时调用的 `cleanup` 方法中，显示了映射中存在的值。

```java
public class DisplayBolt extends BaseRichBolt {
    private static final long serialVersionUID = 1L;
    private HashMap<String, long> counts = null;

    public void prepare(@SuppressWarnings("rawtypes") Map config, TopologyContext context, OutputCollector collector) {
        this.counts = new HashMap<String, long>();
    }

    public void execute(Tuple tuple) {
        String word = tuple.getStringByField("word");
        long count = tuple.getLongByField("count");
        this.counts.put(word, count);
    }

    public void declareOutputFields(OutputFieldsDeclarer declarer) {
        // this bolt does not emit anything
    }

    public void cleanup() {
        System.out.println("--- FINAL COUNTS ---");
        List<String> keys = new ArrayList<String>();
        keys.addAll(this.counts.keySet());
        Collections.sort(keys);
        for (String key : keys) {
            System.out.println(key + " : " + this.counts.get(key));
        }
        System.out.println("--------------");
    }
}
```

- **创建拓扑并提交**：在定义所有 spout 和 bolt 之后，将它们绑定到一个程序（即拓扑）中。这里有两件事非常重要，具有 ID 的 bolt 序列和流的分组。

  首先，在第 1 行创建 `TopologyBuilder`，这是构建 spout 和 bolt 的完整拓扑所必需的。在第 2 行设置一个 spout，即 `FixedSentenceSpout spout`。设置第 3 行中的第一个 bolt，即 `SplitSentenceBolt`。现在我们使用了 `shuffleGrouping`，这意味着所有元组将在所有任务中均匀分布。在第 4 行设置第二个 bolt，即 `WordCountBolt`。在这里使用 `fieldsGrouping`，因为我们希望相同的单词进入同一个进程来执行单词计数。设置最后一个 bolt，即 `DisplayBolt` bolt。一旦拓扑关闭，此 bolt 将显示最终输出。

```
TopologyBuilder builder = new TopologyBuilder(); # line 1
builder.setSpout("sentence-spout", new FixedSentenceSpout()); # line 2
builder.setBolt("split-bolt", new
SplitSentenceBolt()).shuffleGrouping("sentence-spout"); # line 3
builder.setBolt("count-bolt", new
WordCountBolt()).fieldsGrouping("split-bolt", new Fields("word"));
# line 4
builder.setBolt("display-bolt", new
DisplayBolt()).globalGrouping("count-bolt"); # line 5
Config config = new Config();

if (mode.equals("cluster")) {
StormSubmitter.submitTopology("word-count-topology", config,
builder.createTopology()); # line 6
} else {
    LocalCluster cluster = new LocalCluster();
    cluster.submitTopology("word-count-topology", config,
    builder.createTopology()); # line 7
    Thread.sleep(20000);
    cluster.killTopology("word-count-topology");
    cluster.shutdown();
}
```

运行任务

要运行上一个示例有两种方法：一种是本地模式，另一种是集群模式。

1. 本地模式

本地模式表示在本地集群上运行拓扑。用户可以在 Eclipse 中运行它，而无须设置和配置 Storm。要在本地集群中运行它，请右键单击 **BasicStormWordCountExample** 并选择 **Run As | Java Application**。日志将在控制台上开始打印。在 20s 关闭之前，最终输出

将显示在控制台上，如图 4-7 所示。

```
--- FINAL COUNTS ---
4 : 4073
Apache : 8145
Storm : 8145
This : 8146
Very : 4073
basic : 4073
chapter : 4073
count : 4073
engine : 4072
example : 12219
is : 12218
of : 8146
open : 4072
processing : 4072
real : 4072
source : 4072
time : 4072
word : 4073
```

图 4-7

2．集群模式

要在集群模式下运行该示例，请执行以下步骤。

① 转到放置 pom.xml 的项目目录，并使用以下命令进行构建。

```
mvn clean install
```

② 在 Storm 集群上提交拓扑 JAR。

```
./storm jar ~/workspace/Practical-Real-time-
Analytics/chapter4/tarrget/chapter4-0.0.1-SNAPSHOT.jar
com.book.realtime.job.BasicStormWordCountExample cluster
```

③ 检查 UI 上的拓扑状态，如图 4-8 所示。

图 4-8

④ 单击拓扑时，将找到 spout 和 bolt 的详细信息，如图 4-9 所示。

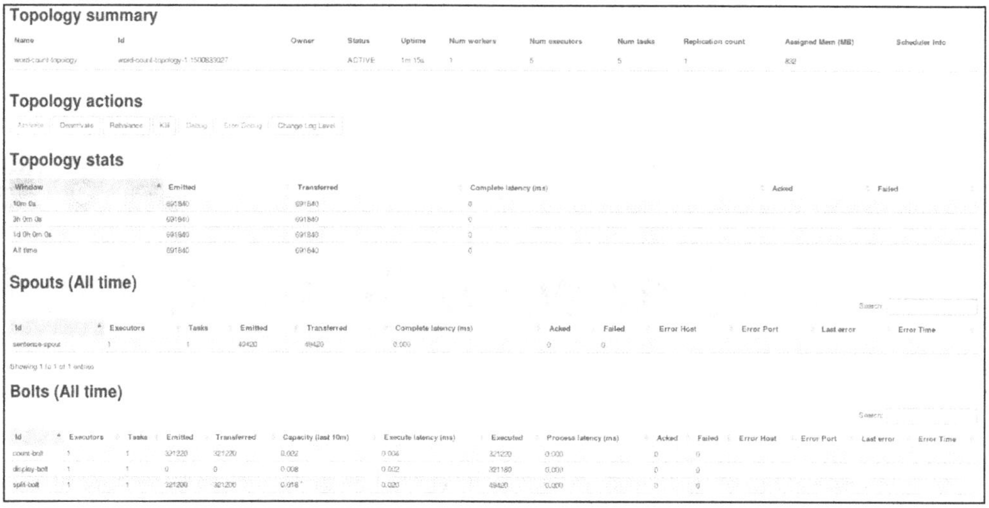

图 4-9

⑤ 从 UI 中杀死拓扑，如图 4-9 所示。

⑥ 检查工作日志以获得最终输出，并查看如下文件。

```
STORM_HOME/logs/workers-artifacts/word-count-topology-1-<Topology ID>/<worker port>/worker.log。
```

4.5 小结

在本章，我们向读者介绍了 Storm 的基础知识。首先介绍 Storm 的历史，让读者了解了 Nathan Marz 是怎样得到 Storm 的想法以及他在将 Storm 作为开源软件发布到 Apache 时面临的挑战；然后讨论了 Storm 及其组件的架构：Nimbus、监管者、执行器和任务都是 Storm 架构的一部分，而 Storm 组件是由元组、流、拓扑、spout 和 bolt 组成的；最后讨论了如何设置 Storm 并将其配置为在集群中运行（集群模式必须首先安装 Zookeeper）。

在本章的最后，我们讨论了使用 spout 和多个 bolt 在 Storm 中实现单词计数的统计示例，展示了如何在本地以及集群上运行示例。

第 5 章
配置 Apache Spark 和 Flink

在本章,我们将帮助读者完成本书所需的各种计算引擎组件的基本安装和配置,带领读者进行组件安装,并结合一些基本示例来验证这些组件。Apache Spark、Apache Flink 和 Apache Beam 是本章中将要讨论的计算引擎。市场上还有很多类似的计算引擎。

根据计算引擎官网上的定义,Apache Spark 是大规模数据处理的快速通用引擎;Apache Flink 是一个开源流处理框架,用于构建分布式、高性能、高可用且精确的数据流应用程序;Apache Beam 是一个开源的统一模型,用于定义批处理和流数据并行处理管道。使用 Apache Beam 时,用户可以在选择的计算引擎(如 Apache Spark、Apache Flink 等)上运行该程序。

—— 本章主要包括以下内容 ——
- 安装并快速运行 Spark
- 安装并快速运行 Flink
- 安装并快速运行 Apache Beam
- Apache Beam 中的平衡

5.1 安装并快速运行 Spark

安装 Spark 有两种不同的方法,分别是利用源代码构建或直接下载安装包解压缩它。下面将介绍这两种不同的安装方法。

5.1.1 源码构建

下载源代码的方式,如图 5-1 所示。

图 5-1

读者将需要使用 Maven 3.3.6 和 Java 7+ 来编译 Spark 2.1.0。此外,用户需要更新环境变量 MAVEN_OPTS,因为默认设置将无法编译代码。

```
export MAVEN_OPTS="-Xmx2g -XX:ReservedCodeCacheSize=512m"
```

使用以下命令进行构建,使用 Hadoop 版本 2.4.0 编译 Spark 2.1.0。

```
./build/mvn -Pyarn -Phadoop-2.4 -Dhadoop.version = 2.4.0 -DskipTests clean package
```

5.1.2 下载 Spark 安装包

使用与上节相同的链接下载最新版本(2.1.0)。如果用户想要安装除最新版本以外的任何内容并选择了预先构建 Hadoop,请选择 Spark 开放版本。用户还可以从同一页面下载源代码并构建它。Spark 2.1.0 版本要求的语言为:Java 7+、Python 3.4+ 和 Scala 2.11。

以下是提取下载的文件 spark-2.1.0-binhadoop2.7.tar 文件。

```
mkdir demo
mv /home/ubuntu/downloads/spark-2.1.0-bin-hadoop2.7.tar~ / demo
cd demo
tar -xvf spark-2.1.0-bin-hadoop2.7.tar
```

解压缩后的文件和文件夹列表,如图 5-2 所示。

其他下载或安装 Spark 的方法是使用 **Cloudera Distribution Hadoop(CDH)**、**Hortonworks Data Platform(HDP)**或 **MapR** 提供的 Spark 虚拟映像来实现的。如果用户希望在 Spark 上获得客户支持,请使用云上的 Databricks。

图 5-2

5.1.3 运行示例

Spark 包附带了示例。因此要测试所有必需的依赖项是否正常工作，请执行以下命令。

```
cd spark-2.1.0-bin-hadoop2.7
./bin/run-example SparkPi 10
```

这个输出将是：

```
Pi is roughly 3.1415431415431416
```

在启动 spark-shell 之前，用户需要在本地机器上设置 Hadoop。下载 Hadoop 2.7.4 版本。使用以下命令解压缩文件。

```
tar -xvf hadoop-2.7.4.tar.gz
```

输出如图 5-3 所示。

图 5-3

将 .bashrc 文件中的配置复制到用户的根目录位置。

```
export HADOOP_PREFIX=/home/impadmin/tools/hadoop-2.7.4
export HADOOP_HOME=/home/impadmin/tools/hadoop-2.7.4
export HADOOP_MAPRED_HOME=${HADOOP_HOME}
export HADOOP_COMMON_HOME=${HADOOP_HOME}
export HADOOP_HDFS_HOME=${HADOOP_HOME}
export YARN_HOME=${HADOOP_HOME}
export HADOOP_CONF_DIR=${HADOOP_HOME}/etc/hadoop
# Native Path
export HADOOP_COMMON_LIB_NATIVE_DIR=${HADOOP_PREFIX}/lib/native
```

```
export HADOOP_OPTS="-Djava.library.path=$HADOOP_PREFIX/lib"
export PATH=$PATH:$HADOOP_HOME/bin:$HADOOP_HOME/sbin
```

使用以下命令启动所有 Hadoop 服务：

```
/bin/start-all.sh
```

输出如图 5-4 所示。

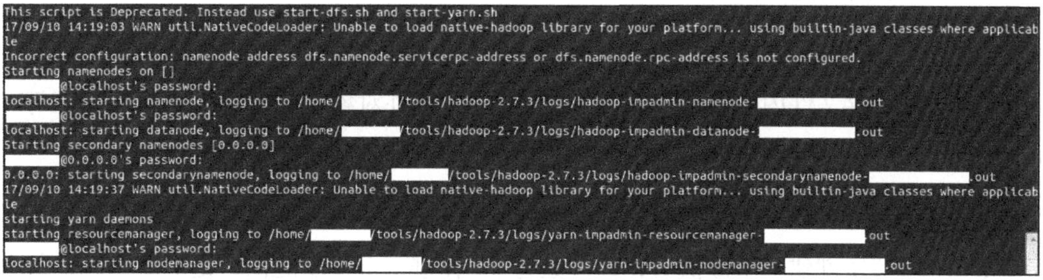

图 5-4

现在，在/tmp 文件夹中创建一个名为 input.txt 的文件，并添加给定的行，如图 5-5 所示。

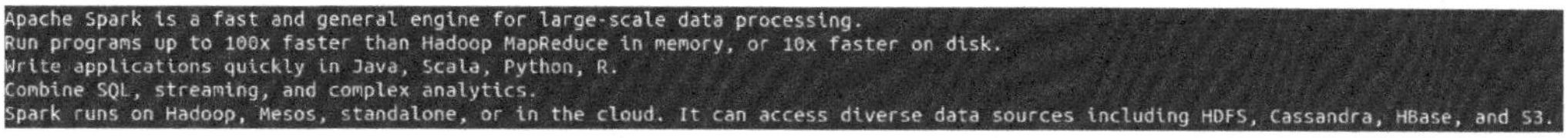

图 5-5

让我们在 Spark 中使用 Scala 创建一个单词计数程序。Spark 提供 shell，在它里面可以编写代码并将其作为命令行执行。

```
cd spark-2.1.0-bin-hadoop2.7
./bin/spark-shell
```

该命令的输出如图 5-6 所示。

图 5-6

按照如下步骤创建单词计数程序。

① 定义文件名。将文件名保存至变量 `fileName` 中。

```
valfileName = ""/tmp/input.txt"
```

② 从变量 `fileName` 中获取文件名。获取文件指针，并使用 Spark 上下文 sc 构建文件对象，用来保存文件内容。

```
valfile = sc.textFile(fileName)
```

③ 使用保存文件内容的 Spark 上下文 sc 构造文件对象，执行字数统计。读取每一行并将每行分隔符拆分为空格" "。

```
valwordCount = file.flatMap(line =>line.split(" "))
```

④ 映射转换创建一个键值对，其计数为 1。在进入下一步之前，键值对将包含每个计数为 1 的单词。

```
map(word => (word, 1))
```

⑤ 将相似的键相加，并给出最终计数值为 1 的单词数。键值对将包含计数为 1 的每个单词。

```
reduceByKey(_ + _)
```

⑥ 收集并打印输出。从所有工作单元中收集包含单词的键值及其计数的值。

```
wordCount.collect()
```

⑦ 在控制台上打印每个键值。

```
foreach(println)
```

⑧ 将输出保存在文件中。

```
wordCount.saveAsTextFile ("/tmp/output")
```

⑨ 输出显示如图 5-7 所示。

完整的程序如下：

```
val fileName = "/tmp/input.txt"
val file = sc.textFile(fileName)
val wordCount = file.flatMap(line=>line.split(" ")).map(word =>(word, 1)).
reduceByKey(_+_)
```

```
wordCount.collect().foreach(println)
wordCount.saveAsTextFile("/tmp/output")
```

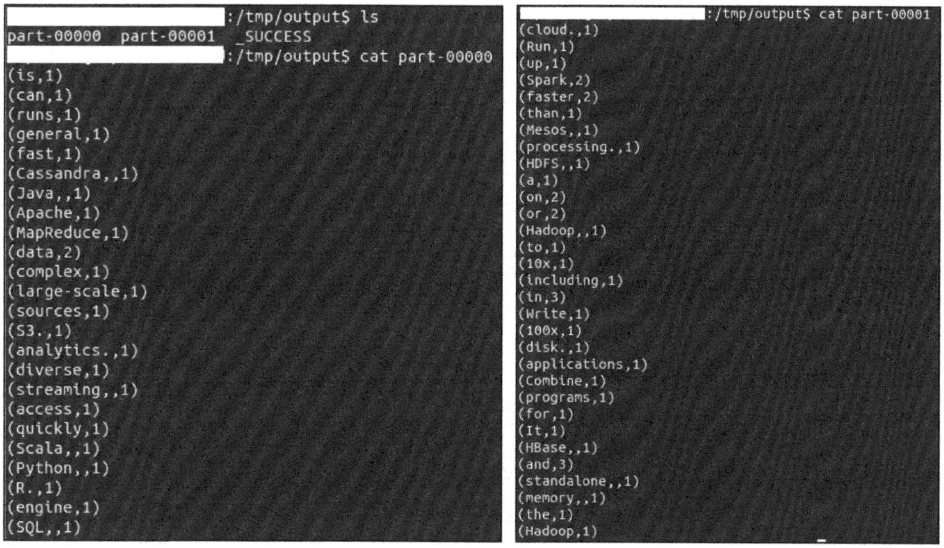

```
:/tmp/output$ ls
part-00000  part-00001  _SUCCESS
            :/tmp/output$ cat part-00000
(is,1)
(can,1)
(runs,1)
(general,1)
(fast,1)
(Cassandra,,1)
(Java,,1)
(Apache,1)
(MapReduce,1)
(data,2)
(complex,1)
(large-scale,1)
(sources,1)
(S3.,1)
(analytics.,1)
(diverse,1)
(streaming,,1)
(access,1)
(quickly,1)
(Scala,,1)
(Python,,1)
(R.,1)
(engine,1)
(SQL,,1)
```

```
            :/tmp/output$ cat part-00001
(cloud.,1)
(Run,1)
(up,1)
(Spark,2)
(faster,2)
(than,1)
(Mesos,,1)
(processing.,1)
(HDFS,,1)
(a,1)
(on,2)
(or,2)
(Hadoop,,1)
(to,1)
(10x,1)
(including,1)
(in,3)
(Write,1)
(100x,1)
(disk.,1)
(applications,1)
(Combine,1)
(programs,1)
(for,1)
(It,1)
(HBase,,1)
(and,3)
(standalone,,1)
(memory,,1)
(the,1)
(Hadoop,1)
```

图 5-7

5.2 安装并快速运行 Flink

安装 Flink 同样也有两种方法：使用源代码构建和下载安装包。下面我们将分别介绍这两种方法。

5.2.1 使用源码构建 Flink

下载源代码或复制 Git 存储库。要从 Git 中复制，请输入以下命令：

```
git clone https://github.com/apache/Flink
```

构建 Flink 需要使用 Maven 3.0.3 和 Java 8。用 Maven 构建 Flink 的命令如下：

```
mvn clean install -DskipTests
```

如果想用不同版本的 Hadoop 构建 Flink，那么使用以下命令：

```
mvn clean install -DskipTests -Dhadoop.version = 2.6.1
```

5.2.2　下载 Flink

下载最新版本的 Flink（1.1.4），如图 5-8 所示。

Binaries

	Scala 2.10	Scala 2.11
Hadoop® 1.2.1	Download	
Hadoop® 2.3.0	Download	Download
Hadoop® 2.4.1	Download	Download
Hadoop® 2.6.0	Download	Download
Hadoop® 2.7.0	Download	Download

图 5-8

用以下命令提取下载的 `Flink-1.1.4-bin-hadoop27-scala_2.11.tgz` 文件：

```
mkdir demo
mv /home/ubuntu/downloads/Flink-1.1.4-bin-hadoop27-scala_2.11.tgz ~/demo
cd demo
tar -xvfFlink-1.1.4-bin-hadoop27-scala_2.11.tgz
```

解压缩后的文件列表如图 5-9 所示。

```
README.txt
log
bin
tools
resources
NOTICE
LICENSE
lib
examples
conf
```

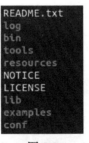

图 5-9

用以下命令启动 Flink：

```
./bin/start-local.sh
```

Apache Flink 提供了一个仪表板，显示正在运行或已完成的所有作业并提交新作业。如果仪表板可以访问 `http://localhost:8081`，则表明一切都已正常启动并运行，如图 5-10 所示。

图 5-10

5.2.3 运行示例

Flink 提供了一个名为 `Flink DataStream API` 的流 API，可以实时处理连续的无界数据流。

要开始使用 `Datastream API`，读者应该将以下依赖项添加到项目中。在这里，我们使用 sbt 进行构建管理。

```
org.apache.Flink"%%"Flink-scala"%"1.0.0
```

在接下来的几个步骤中，将创建一个单词计数统计程序，从套接字上读取并实时显示字数。

① 必须搭建程序运行的流式环境。我们将在本章后面讨论部署模式。代码如下：

```
val environment = StreamExecutionEnvironment.getExecutionEnvironment
```

 `StreamExecutionEnvironment` 类似于 Spark 上下文。

② 导入 Java 或 Scala 的流 API。

```
import org.apache.Flink.streaming.api.scala._
```

③ 从套接字中创建 `DataStream`。

```
valsocketStream = environment.socketTextStream("localhost", 9000)
```

`socketStream` 的类型为 `DataStream`。`DataStream` 是 Flink 的流 API 的基本抽象。

④ 实现 wordcount 逻辑。

```
valwordsStream = socketStream.flatMap(value =>value.split("\\s+")).map(value =>
(value,valkeyValuePair = wordsStream.keyBy(0)
valcountPair = keyValuePair.sum(1)
```

keyBy 函数与 the groupBy 函数相同，sum 函数与 reduce 函数相同。keyBy 和 sum 中的 0 和 1 分别表示元组中列的索引。

⑤ 打印字数。

```
countPair.print()
```

⑥ 触发程序执行。用户需要显式调用 execute 函数来触发执行。代码如下：

```
env.execute()
```

完整代码如下：

```
importorg.apache.Flink.streaming.api.scala._
objectStreamingWordCount {
def main(args: Array[String]) {
valenv = StreamExecutionEnvironment.getExecutionEnvironment

    // create a stream using socket
valsocketStream = env.socketTextStream("localhost",9000)

    // implement word count
valwordsStream = socketStream.flatMap(value
=>value.split("\s+")).map(value => (value,1))
valkeyValuePair = wordsStream.keyBy(0)
valcountPair = keyValuePair.sum(1)

    // print the results
countPair.print()

    // execute the program
env.execute()
  }
}
```

读者可以使用 sbt 构建上述程序并创建 .jar 文件。单词计数统计程序是预先构建的，并附带 Flink 安装包。读者可以在以下位置找到 JAR 文件。

```
~/demo/Flink-1.1.4/examples/streaming/SocketWindowWordCount.jar
```

5.2 安装并快速运行 Flink

用以下命令在 9000 端口启动套接字：

```
nc -lk 9000
```

以下命令将在 Flink 上提交作业（替换读者的 JAR 文件名）。

```
/demo/ Flink-1.1.4/bin/Flink run
examples/streaming/SocketWindowWordCount.jar --port 9000
```

Flink 仪表板开始显示正在运行的作业及其相关详细信息，如图 5-11 所示。

图 5-11

图 5-12 显示了每个子任务的作业详细信息。

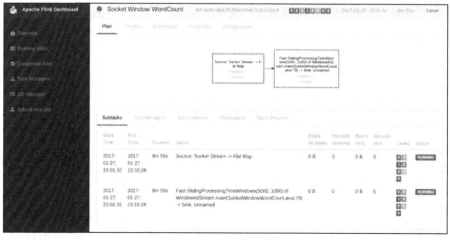

图 5-12

现在，在控制台上输入单词或语句，如图 5-13 所示。

图 5-13

输出如图 5-14 所示。

图 5-14

当市场上有多种可用于流式传输的选项时，采用哪种技术取决于性能、提供的功能、可靠性以及它如何适用例等因素。表 5-1 给出了 Storm、Spark 和 Flink 之间的一些比较。

表 5-1

	Storm	Spark	Flink
版本	1.0	2.1.0	1.1.4
有状态处理	Storm 1.0 通过 Redis 后端引入状态处理。状态保持在 bolt 实例水平。存储是非分布式的	通过 key 更新状态	支持，分区状态存储（由文件系统或者 RocksDB）
窗口机制	根据持续时间或事件计数滑动和翻滚窗口	滑动窗口	滑动窗口、翻滚窗口、自定义窗口、基于事件计数的窗口

续表

	Storm	Spark	Flink
任务隔离	任务在执行中不是孤立的。多个任务可以在单个 JVM 中执行，任务可能受其他任务影响	独立部署，或者通过 Yarn 或者 Mesos 实现任务隔离	任务运行在 Yarn 容器中
资源管理	Nimbus	独立部署，或者通过 Yarn 或者 Mesos 实现任务隔离	独立部署、或通过 Yarn
SQL 兼容性	不兼容	不兼容	支持 Table API 和 SQL，但是 SQL 还不够成熟
开发语言支持	Java、JavaScript、Python、Ruby	Java、Scala、Python	Java、Scala

5.3 安装并快速运行 Apache Beam

什么是 **Apache Beam**？根据其官网的定义，Apache Beam 是一种统一的编程模型，允许用户在任何执行引擎上运行批处理和流处理作业。

为何选择 Apache Beam？出于以下几点考虑。

- **统一性**：为批处理和流用例统一编程模型。
- **可移植性**：运行时环境与代码解耦。可以在多个运行环境中执行管道，包括 Apache Apex、Apache Flink、Apache Spark 和 Google Cloud Dataflow。
- **可扩展性**：编写和共享新的 SDK，I/O 连接器和转换库。用户可以创建自己的运行器以便支持新的运行环境。

5.3.1 Beam 模型

在 Beam 中执行的任何变换或聚合称为 `PTransform`，这些变换之间的连接称为 `PCollection`。

`PCollection` 可以是有界或无界的。一组或多组 `PTransform` 和 `PCollection` 在 Beam 中构成管道，如图 5-15 所示。

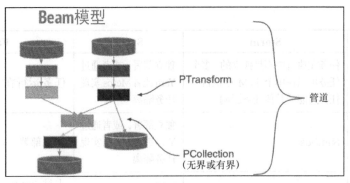

图 5-15

5.3.2 运行示例

用户需要 Java 7 或更高版本以及 Maven 来构建代码。接下来用 Apache Beam 运行著名的单词计数示例。

前面的 Maven 命令生成一个包含 Apache Beam 的 **WordCount** 示例的 Maven 项目。

```
$ mvnarchetype:generate \
    -DarchetypeRepository=https://repository.apache.org/content
     /groups/snapshots \
    -DarchetypeGroupId=org.apache.beam \
    -DarchetypeArtifactId=beam-sdks-java-maven-archetypes-examples \
    -DarchetypeVersion=LATEST \
    -DgroupId=org.example \
    -DartifactId=word-count-beam \
    -Dversion="0.1" \
    -Dpackage=org.apache.beam.examples \
    -DinteractiveMode=false
```

这将创建一个名为 `word-count-beam` 的文件夹，其中包含以下代码。

```
$ cd word-count-beam/
$ ls
pom.xml src
$ lssrc/main/java/org/apache/beam/examples/
DebuggingWordCount.java    WindowedWordCount.java common
MinimalWordCount.java              WordCount.java
```

Apache Beam 提供了不同版本的 `WordCount` 作为示例。这里只关注 `WordCount.java`。Beam 管道可以运行多个运行器。在执行任何管道时，用户必须使用参数 `--runner =`

`<runner>`指定运行器。

要运行 `WordCount` 示例,请执行指定运行器对应的命令。

- **Direct 运行器**:没有必要指定运行器,因为它是默认的。Direct 运行器在本地计算机上运行,不需要特别设置。命令如下:

```
mvn compile exec:java -
Dexec.mainClass=org.apache.beam.examples.WordCount -Dexec.args="--
inputFile=pom.xml --output=counts" -Pdirect-runner
```

- **Flink 运行器**的命令如下:

```
mvn compile exec:java -
Dexec.mainClass=org.apache.beam.examples.WordCount -Dexec.args="--
runner=FlinkRunner --inputFile=pom.xml --output=counts" -PFlinkrunner
```

- **Spark 运行器**的命令如下:

```
mvn compile exec:java -
Dexec.mainClass=org.apache.beam.examples.WordCount -Dexec.args="--
runner=SparkRunner --inputFile=pom.xml --output=counts" -Psparkrunner
```

- **DataFlow 运行器**的命令如下:

```
mvn compile exec:java -
Dexec.mainClass=org.apache.beam.examples.WordCount -Dexec.args="--
runner=DataflowRunner --gcpTempLocation=gs://<your-gcs-bucket>/tmp
--inputFile=gs://apache-beam-samples/shakespeare/* --
output=gs://<your-gcs-bucket>/counts" -Pdataflow-runner
```

运行上一个命令后,将在同一文件夹中创建以 count 开头的文件名。当我们执行命令来检查文件中的条目时,输出如图 5-16 所示。

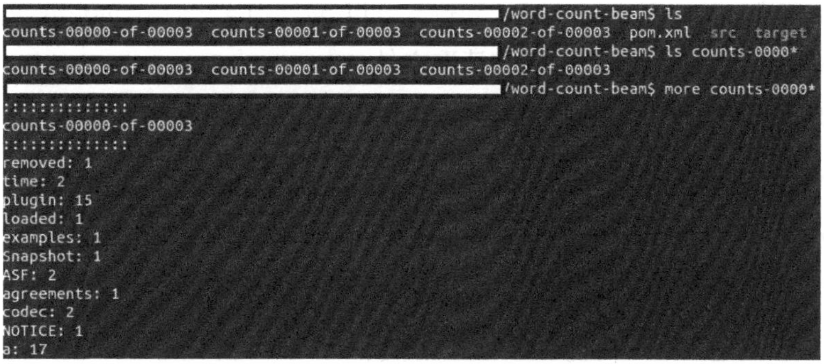

图 5-16

如果读者的程序由于缺少 Maven 中的依赖项而失败,那么请清理一下文件夹 .m2/repository/org/apache/maven 并再试一次。

5.3.3 MinimalWordCount 示例

在本节,我们将介绍 Apache Beam 站点上给出的 MinimalWordCount.java 示例,并在前面设置的 Flink 服务器上运行它。在上面示例中运行的 WordCount 与 MinimalWordCount.java 具有相同的实现,但具有最佳编码实践和可重用代码。因此,在这里将讨论 MinimalWordCount 示例,以便用户可以清楚地理解该概念。在 Beam 管道中构建最小计数示例的关键概念包括创建管道、应用管道转换、计数转换和运行管道等。

- **创建管道**的命令如下:

```
PipelineOptions options = PipelineOptionsFactory.create();
```

PipelineOptions 对象包含有关运行器和特定运行器配置的信息。默认情况下,它在默认运行器上运行。若使其特定于 Flink 或 Spark 运行器,请进行特定于运行器的更改。

```
FlinkPipelineOptions options =
PipelineOptionsFactory.create().as(FlinkPipelineOptions.class);
options.setRunner(FlinkRunner.class);
```

使用在上面定义的选项创建 Pipeline 对象。

```
Pipeline p = Pipeline.create(options);
```

- **应用管道转换**:每个转换都接受输入并以 PCollection 的形式生成输出。应用读取转换,设置输入路径并读取文件。

```
p.apply(TextIO.Read.from("gs://apache-beam-samples/shakespeare/*"))
```

gs:// 表示 Google 云端存储。开发者可以使用完整路径指定本地文件,但请记住,如果在 Spark 集群或 Flink 集群上运行代码,则该文件可能不存在于 Readfrom 函数提到的目录中。

调用 DoFn 实施 ParDo 转换。DoFn 对来自 TextIO 的 PCollection 进行元素拆分,并生成一个新的 PCollection,每个单词作为一个元素。命令如下:

```
.apply("ExtractWords", ParDo.of(new DoFn<String, String>() {
        @ProcessElement
public void processElement(ProcessContext c) {
for (String word : c.element().split("[^a-zA-Z']+")) {
if (!word.isEmpty()) {
c.output(word);
        }
      }
    }
  }))
```

计数转换，计算 PCollection 中每个元素的出现次数，并生成键值对形式的 PCollection。每个键代表一个唯一的单词，值是总计数。命令如下：

```
.apply(Count.<String>perElement())
```

要将输出转换为可读/有意义的格式，请应用 MapElement 复合变换，该变换为 PCollection 中的每个元素执行用户定义的函数，并生成字符串或任何类型的 PCollection。在这里，它以可打印的格式转换键值对。

```
.apply("FormatResults", MapElements.via(new
SimpleFunction<KV<String, Long>, String>() {
        @Override
public String apply(KV<String, Long> input) {
returninput.getKey() + ": " + input.getValue();
        }
      }))
```

要将输出保存在文件中，应用写入转换，该转换需要 PCollection 并生成 PDone。命令如下：

```
.apply(TextIO.Write.to("wordcounts"));
```

- **运行管道**：在代码中使用以下语句运行管道。

```
p.run().waitUntilFinish();
```

管道运行流程，如图 5-17 所示。

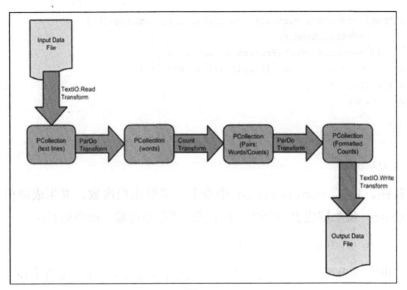

图 5-17

完整的程序如下：

```java
packageorg.apache.beam.examples;

importorg.apache.beam.runners.Flink.FlinkPipelineOptions;
importorg.apache.beam.runners.Flink.FlinkRunner;
importorg.apache.beam.sdk.Pipeline;
importorg.apache.beam.sdk.io.TextIO;
importorg.apache.beam.sdk.options.PipelineOptionsFactory;
importorg.apache.beam.sdk.transforms.Count;
importorg.apache.beam.sdk.transforms.DoFn;
importorg.apache.beam.sdk.transforms.MapElements;
importorg.apache.beam.sdk.transforms.ParDo;
importorg.apache.beam.sdk.transforms.SimpleFunction;
importorg.apache.beam.sdk.values.KV;

public class MinimalWordCount {

public static void main(String[] args) {
FlinkPipelineOptions options =
PipelineOptionsFactory.create().as(FlinkPipelineOptions.class);
    options.setRunner(FlinkRunner.class);
```

```java
    Pipeline p = Pipeline.create(options);
p.apply(TextIO.Read.from("gs://apache-beam-samples/shakespeare/*"))
    .apply("ExtractWords", ParDo.of(new DoFn<String, String>() {
                    @ProcessElement
public void processElement(ProcessContext c) {
for (String word : c.element().split("[^a-zA-Z']+")) {
if (!word.isEmpty()) {
c.output(word);
                    }
                   }
                  }
                 }))
    .apply(Count.<String>perElement())
     .apply("FormatResults", MapElements.via(new SimpleFunction<KV<String, Long>,
String>() {
                    @Override
public String apply(KV<String, Long> input) {
return input.getKey() + ": " + input.getValue();
                   }
                  }))
    .apply(TextIO.Write.to("wordcounts"));
p.run().waitUntilFinish();
  }
}
```

5.4 Apache Beam 中的平衡

Apache Beam 提供一种在完整性、延迟和成本之间保持平衡的方法。在这里，完整性是指所有事件应该如何处理，延迟是指执行一个事件所花费的时间，成本是指完成任务所需的计算能力。以下是在 Apache Beam 中构建管道时应提出的问题，该管道可在上述 3 个参数之间保持平衡。

- **计算结果是什么**？使用管道中可用的转换，系统正在计算结果。
- **在哪里计算事件中的时间结果**？这是通过使用事件时间窗口来实现的。事件时间窗口进一步分为固定窗口、滑动窗口和会话窗口。
- **在处理时间内何时实现了结果**？这是通过使用水印和触发器实现的。水印是衡量无界流中一系列事件完整性的方法。触发器定义何时从时间窗口发出输出。这些是实现平衡的重要因素。

- **结果的改进如何相关**？累加器用于细化上述过程产生的结果。

上面的转换示例计算了一批文件的字数。接下来将继续使用窗口、水印、触发器和累加器。最后将讨论随着 Apache Beam 示例给出的示例。

示例说明了触发器的不同情况，其中部分生成结果，包括通过重新计算结果来延迟到达的数据。数据是来自圣地亚哥的实时流量，由沿着每条高速公路设置的传感器站传来。每个传感器读数包括计算该高速公路上所有车道的总流量。输入将是一个文本文件，输出将写入 Big Query。表 5-2 所示为一系列记录。

表 5-2

键(freeway)	值(total_flow)	事件时间	处理时间
5	50	10:00:03	10:00:47
5	30	10:01:00	10:01:03
5	30	10:02:00	11:07:00
5	20	10:04:10	10:05:15
5	60	10:05:00	11:03:00
5	20	10:05:01	11:07:30
5	60	10:15:00	10:27:15
5	40	10:26:40	10:26:43
5	60	10:27:20	10:27:25
5	60	10:29:00	11:11:00

设置窗口持续时间为 5min。命令如下：

```
PCollection<TableRow>defaultTriggerResults = flowInfo
.apply("Default", Window.<KV<String,
Integer>>into(FixedWindows.of(Duration.standardMinutes(30))))
```

当系统的水印通过窗口结束时，触发器会发出以下输出：

```
.triggering(Repeatedly.forever(AfterWatermark.pastEndOfWindow()))
```

在水印通过到达元素的事件时间戳之后，到达的数据被视为延迟数据。示例中，如果事件迟到，则该事件会被丢弃。

```
.withAllowedLateness(Duration.ZERO)
```

窗口完成后元素被丢弃,它不会被转移到下一个窗口。

```
.discardingFiredPanes()
```

结果见表 5-3。

表 5-3

键 (freeway)	值(total_flow)	number_of_records	isFirst	isFirst	Timing
5	260	6	真	真	ON_TIME

默认触发器生成的每个窗格(不允许有延迟)都将是窗口中的第一个也是最后一个窗格,并将处于 ON_TIME 状态。在 11∶03∶00(处理时间)时,系统水印可能已提升到 10∶54∶00。因此,当事件时间为在 10∶05∶00 时的数据记录但在 11∶03∶00 到达时,它被视为延迟并被丢弃。

示例中,用户可以更改允许延迟到达的数据的持续时间,如下所示:

```
.withAllowedLateness(Duration.standardDays(1)))
```

结果见表 5-4。

表 5-4

键 (freeway)	值 (total_flow)	number_of_records	isFirst	isFirst	timing
5	260	6	真	真	及时
5	60	1	假	假	延迟
5	30	1	假	假	延迟
5	20	1	假	假	延迟
5	60	1	假	假	延迟

这导致在水印通过窗口结束后,每个窗口都为 ONE_DAY,保持打开。如果用户希望跨窗格累加该值,并且还希望在不考虑水印的情况下发射结果,则可以在以下示例中实现代码。

每当在窗格中接收到元素时,触发器都会在处理时间后发出该元素。

```
triggering(Repeatedly.forever(AfterProcessingTime.pastFirstElementInPane()
```

元素将会被累积，以便每个近似值除了新到达的数据，还包括所有以前的数据。

`accumulatingFiredPanes()`

结果见表 5-5。

表 5-5

键 (freeway)	值 (total_flow)	number_of_records	isFirst	isFirst	timing
5	80	2	真	假	早
5	100	3	假	假	早
5	260	6	假	假	早
5	320	7	假	假	延迟
5	370	9	假	假	延迟
5	430	10	假	假	延迟

由于没有任何依赖于水印的触发器，因此不会触发 `ON_TIME`。相反，所有窗格都是 `EARLY` 或 `LATE`。

5.5 小结

在本章，我们旨在让读者了解 Spark、Flink 和 Beam 的安装配置和快速执行。通过这些组件，读者可以在单机模式和集群模式下轻松地运行 jar 中捆绑代码对应的运行示例。

Storm 也是一种计算引擎，我们将在接下来的几章中讨论 Storm。在第 6 章，我们将讨论 Storm 与不同数据源的集成。

第三部分　Storm 实时计算

- 第 6 章　集成 Storm 与数据源
- 第 7 章　从 Storm 到 Sink
- 第 8 章　Storm Trident

第 6 章
集成 Storm 与数据源

在本章，我们将数据源与 Storm 分布式计算引擎进行集成，涉及将流数据的源连接到 RabbitMQ 之类的代理服务，然后将流管道连接到 Storm。有一个非常有趣的传感器数据案例，它从免费的实时传感器数据通道采集实时数据并使用流式传输，然后将其推送到 RabbitMQ，然后将其推送到 Storm 拓扑以进行业务分析。

---- 本章主要包括以下内容 ----

- RabbitMQ 有效的消息传递
- RabbitMQ 交换器
- RabbitMQ 安装配置
- RabbitMQ 的发布和订阅
- RabbitMQ 与 Storm 集成
- PubNub 数据流发布者
- 传感器数据处理拓扑

6.1 RabbitMQ 有效的消息传递

RabbitMQ 是 Storm 生产实现中最受欢迎的代理/队列服务之一，是一个非常健壮和通用的消息传递系统，支持开源和所有主流商业操作系统。队列具有持久化配置和在内存配置两种方式，因此开发人员可以灵活地在可靠性和性能之间做出权衡。

以下是在 RabbitMQ 或其他队列系统中经常使用的几个术语。

- **生产者（Producer）/发布者（Publisher）**：是将消息写入或发送到队列的客户端组件。
- **队列（Queue）**：存储消息的内存缓冲区，是消息被发送到队列的时间到消费者应用程序从队列中读取消息的时间。
- **消费者（Consumer）/订阅者（subscriber）**：是接收或读取队列中消息的客户端组件。

RabbitMQ 的生产者/发布者从未向队列发布任何消息，实际上是将消息写入 **RabbitMQ** 交换器，而交换器又根据交换类型和路由密钥进一步将消息推送到队列中。

6.2 RabbitMQ 交换器

RabbitMQ 具有多功能，提供了多种交换器，以供开发人员使用，来解决开发过程中遇到的各种问题。

6.2.1 直接交换器

在这种类型的交换器中，用户有一个绑定到队列的路由密钥，并将它作为将消息定向到队列的传递密钥。因此，每个发布到交换器的消息都有一个与其关联的路由密钥，这决定了交换器写入的目标队列。例如，在图 6-1 中，消息被写入绿色队列，因为消息路由队列**绿色**绑定到绿色队列。

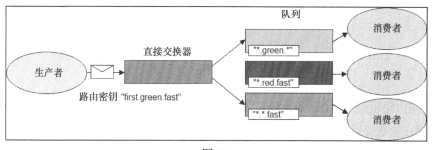

图 6-1

1. 扇出交换器

扇出交换器也可以称为广播交换器，因为当消息发布到扇出交换器时，它被写入/发送到绑定了交换器的所有队列。扇出交换器的工作原理如图 6-2 所示。在这里，生产者发布的消息被发送到所有 3 个队列：green、red 和 orange。因此，简而言之，绑定到交换器的每个队列都会收到一条消息,这类似于发布-订阅代理模式。

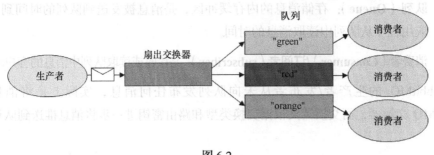

图 6-2

2. 主题交换器

当消息发布到主题交换器时,它将被发送到与路由密钥相匹配的所有队列,或者已发布消息的路由密钥的一部分。例如,如果我们使用密钥向主题交换器发布消息,如果密钥是"green.red",那么消息将被发布到绿色队列和红色队列,如图 6-3 所示。为了帮助读者更好地理解,这里采用比喻解释。

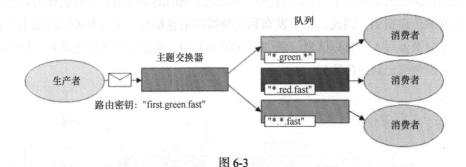

图 6-3

消息路由密钥是"**first.green.fast**",并且在主题交换器中被发布到"green "队列,因为green 出现在主题交换器中。"**red.fast**"队列因为 fast 出现在那里,并且"***.fast**"队列因为"fast"字出现在那里。

3. 标题交换器

标题交换器实际上是通过将消息头与绑定队列头相匹配来将消息发布到特定队列的。这些交换器与基于主题的交换器非常相似,但概念有所不同,即它们具有多个复杂有序条件的匹配标准,如图 6-4 所示。

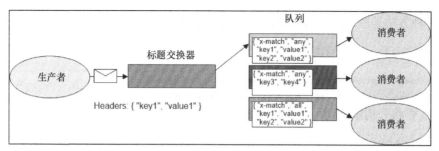

图 6-4

在前一个交换器中发布的消息有一个与之关联的密钥（"**key1**"），它映射到值（"**value1**"），交换器将它与队列头中的所有绑定进行匹配，并且标准实际上与第一个队列中的第一个头值相匹配。第一个队列映射到"**key1**"和"**value1**"，因此消息只发布到第一个队列，同时注意匹配标准为"**any**"。在最后一个队列中，值与第一个队列相同，但匹配标准为"**all**"，这意味着两个映射都应匹配，因此消息不会在底部队列中发布。

6.2.2 RabbitMQ 安装配置

接下来开始安装 RabbitMQ 并查看它的一些行为。从官网上下载最新版本的 RabbitMQ，保存为 DEB 文件，用 Ubuntu 软件中心进行安装，如图 6-5 所示。

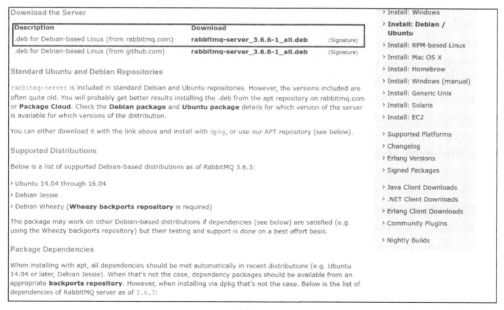

图 6-5

另一种方法是使用命令行界面，使用以下步骤来安装。

① 给定命令用于将资源库添加到读者的系统。

```
echo 'deb http://www.rabbitmq.com/debian/ testing main' |
     sudo tee /etc/apt/sources.list.d/rabbitmq.list
```

② 需要将公钥添加到可信密钥配置中。

```
wget -O- https://www.rabbitmq.com/rabbitmq-release-signing-key.asc
|       sudo apt-key add -
```

③ 成功执行前面的步骤后，更新包列表。

```
sudo apt-get update
```

④ 准备安装 RabbitMQ 服务器。

```
sudo apt-get install rabbitmq-server
```

⑤ 设置完成后，启动服务器。

```
sudo service rabbitmq-server start
```

⑥ 以下命令用于检查服务器的状态：

```
sudo service rabbitmq-server status
```

⑦ 以下命令用于停止 RabbitMQ 服务器：

```
sudo service rabbitmq-server stop
```

⑧ 用户可以/应该使用给定的命令启用 RabbitMQ 管理控制台以提供对 UI 应用程序的访问。

```
sudo rabbitmq-plugins enable rabbitmq_management
```

⑨ 访问 RabbitMQ UI。

```
http://localhost:15672/#/
```

成功完成前面的步骤后，结果如图 6-6 所示。

图 6-6

6.2.3 RabbitMQ 的发布和订阅

安装好 RabbitMQ 并且已经启动和运行后，用户可以通过 RMQ 来验证服务正常与否。显而易见，用户接下来就可以在管理控制台下快速编写一个发布者-消费者应用程序。生产者的代码片段如下所示，该生产者通过一个构建在 MYQueue 上的 MYExchange 交换器来发布消息。

```
package com.book.rmq;
...
public class RMQProducer {
    private static String myRecord;
    private static final String EXCHANGE_NAME = "MYExchange";
    private final static String QUEUE_NAME = "MYQueue";
    private final static String ROUTING_KEY = "MYQueue";
    public static void main(String[] argv) throws Exception {
        ConnectionFactory factory = new ConnectionFactory();
```

第 6 章 集成 Storm 与数据源

```
            Address[] addressArr = { new Address("localhost", 5672) };
            Connection connection = factory.newConnection(addressArr);
            Channel channel = connection.createChannel();
            channel.exchangeDeclare(EXCHANGE_NAME, "direct");
            channel.queueDeclare(QUEUE_NAME, true, false, false, null);
            channel.queueBind(QUEUE_NAME, EXCHANGE_NAME, ROUTING_KEY);
            int i = 0;
              while (i < 1) {
                  try {
                      myRecord = "My Sample record";
                      channel.basicPublish(EXCHANGE_NAME, ROUTING_KEY,
MessageProperties.PERSISTENT_TEXT_PLAIN,
                                  myRecord.getBytes());
                      System.out.println(" [x] Sent '" + myRecord + "'
                                  sent at " + new Date());
                      i++;
                      Thread.sleep(2);
                  } catch (Exception e) {
                      e.printStackTrace();
                  }
              }
              channel.close();
              connection.close();
        }
}
```

接下来的操作如下。

- 将交换器声明为 `MYExchange`，使用路由队列 `MYQueue` 将 `MYQueue` 队列绑定到交换器。

- 接下来声明一个名为 `factory` 的 `ConnectFactory`，并将其绑定到在 `localhost` 上运行的 RabbitMQ 代理中（读者也可以指定相同的本地或远程 IP）。

- 使用以下语句来声明直接交换器：

```
channel.exchangeDeclare(EXCHANGE_NAME, "direct");
```

- 接着声明一个通道，将交换器和队列关联到通道，并使用路由密钥将队列绑定到交换器。

- 创建一个测试消息并使用 `basicPublish` 方法将其发布到队列。

执行程序时的命令行和 UI 输出如图 6-7 所示。

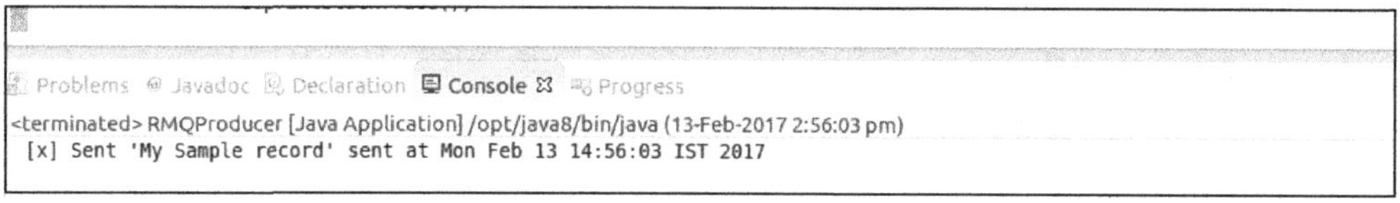

6.2 RabbitMQ 交换器

```
Problems  Javadoc  Declaration  Console  Progress
<terminated> RMQProducer [Java Application] /opt/java8/bin/java (13-Feb-2017 2:56:03 pm)
 [x] Sent 'My Sample record' sent at Mon Feb 13 14:56:03 IST 2017
```

图 6-7

在 RabbitMQ 控制台上，用户可以在 **Queues** 选项卡的 MYQueue 下看到发布的事件，如图 6-8 所示。

图 6-8

接下来组建一个快速的消费者 Java 应用程序，以从队列中读取此消息。

```java
package com.book.rmq;
..
public class RMQConsumer {
    private static final String EXCHANGE_NAME = "MYExchange";
    private final static String QUEUE_NAME = "MYQueue";
    private final static String ROUTING_KEY = "MYQueue";
    public static void main(String[] argv) {
        ConnectionFactory factory = new ConnectionFactory();
        Address[] addressArr = { new Address("localhost", 5672) };
        try {
```

```java
            Connection connection = factory.newConnection(addressArr);
            Channel channel = connection.createChannel();
            connection = factory.newConnection();
            channel.exchangeDeclare(EXCHANGE_NAME, "direct");
            channel.queueDeclare(QUEUE_NAME, true, false, false, null);
            channel.queueBind(QUEUE_NAME, EXCHANGE_NAME, ROUTING_KEY);
            System.out.println("{N|T} Waiting for messages.");
            channel.queueDeclare(QUEUE_NAME, true, false, false, null);
            channel.queueBind(QUEUE_NAME, EXCHANGE_NAME, ROUTING_KEY);
            Consumer consumer = new DefaultConsumer(channel) {
                @Override
                public void handleDelivery(String consumerTag, Envelope
                envelope, AMQP.BasicProperties properties,
                byte[] body) throws IOException {
                        String message = new String(body, "UTF-8");
                        System.out.println("Java Queue - Message Received '" +
                        message + "'");
                }
            };
            // loop that waits for message
            channel.basicConsume(QUEUE_NAME, true, consumer);
        } catch (IOException e) {
            System.out.println("RabbitMQ server is Down !");
            System.out.println(e.getMessage());
        } catch (TimeoutException e) {
            e.printStackTrace();
        }
    }
}
```

读取示例的控制台输出，如图 6-9 所示。

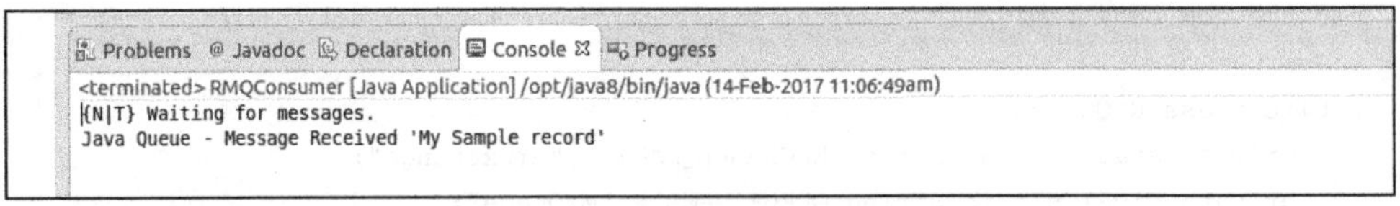

```
Problems  @ Javadoc  Declaration  Console ⌧  Progress
<terminated> RMQConsumer [Java Application] /opt/java8/bin/java (14-Feb-2017 11:06:49am)
{N|T} Waiting for messages.
Java Queue - Message Received 'My Sample record'
```

图 6-9

下一个项目是使用 Storm 拓扑从 RabbitMQ 中读取消息。

6.3 RabbitMQ 与 Storm 集成

截至目前，我们已经完成了 RabbitMQ 基本的安装、发布和订阅。接下来，介绍 RabbitMQ 与 Storm 的集成，并为读者演示一个端到端用例。

AMQPSpout

Storm 通过 `AMQPSpout` 与 RabbitMQ 集成，后者从 RabbitMQ 中读取消息并将其推送到 Storm 拓扑以进行进一步处理。`AMQPSpout` 的编码关键点如下代码片段所示。

```java
..
public class AMQPSpout implements IRichSpout {
    private static final long serialVersionUID = 1L;
    /**
     * Logger instance
     */
    private static final Logger log =
    LoggerFactory.getLogger(AMQPSpout.class);
    private static final long CONFIG_PREFETCH_COUNT = 0;
    private static final long DEFAULT_PREFETCH_COUNT = 0;
    private static final long WAIT_AFTER_SHUTDOWN_SIGNAL = 0;
    private static final long WAIT_FOR_NEXT_MESSAGE = 1L;

    private static final String EXCHANGE_NAME = "MYExchange";
    private static final String QUEUE_NAME = "MYQueue";
    private String amqpHost;
    private int amqpPort;
    private String amqpUsername;
    private String amqpPasswd;
    private String amqpVhost;
    private Boolean requeueOnFail;
    private Boolean autoAck;

    private int prefetchCount;

    private SpoutOutputCollector collector;

    private Connection amqpConnection;
    private Channel amqpChannel;
    private QueueingConsumer amqpConsumer;
    private String amqpConsumerTag;
```

```java
        private Boolean spoutActive;

    // The constructor where we set initialize all properties
    public AMQPSpout(String host, int port, String username, String
        password, String vhost, Boolean requeueOnFail, Boolean autoAck) {
        this.amqpHost = host;
        this.amqpPort = port;
        this.amqpUsername = username;
        this.amqpPasswd = password;
        this.amqpVhost = vhost;
        this.requeueOnFail = requeueOnFail;
        this.autoAck = autoAck;
    }

/*
 * Open method of the spout , here we initialize the prefetch count, this
 * parameter specified how many messages would be prefetched from the queue
 * by the spout - to increase the efficiency of the solution
 */
public void open(@SuppressWarnings("rawtypes") Map conf,
        TopologyContext context, SpoutOutputCollector collector) {
    long prefetchCount = (long) conf.get(CONFIG_PREFETCH_COUNT);
    if (prefetchCount == null) {
            log.info("Using default prefetch-count");
            prefetchCount = DEFAULT_PREFETCH_COUNT;
    } else if (prefetchCount < 1) {
            throw new IllegalArgumentException(CONFIG_PREFETCH_COUNT
                    + " must be at least 1");
    }
    this.prefetchCount = prefetchCount.intValue();

    try {
            this.collector = collector;

            setupAMQP();
    } catch (IOException e) {
        log.error("AMQP setup failed", e);
        log.warn("AMQP setup failed, will attempt to
        reconnect...");
        Utils.sleep(WAIT_AFTER_SHUTDOWN_SIGNAL);
        try {
                reconnect();
```

```java
                } catch (TimeoutException e1) {
                    // TODO Auto-generated catch block
                    e1.printStackTrace();
                }
            } catch (TimeoutException e) {
                // TODO Auto-generated catch block
                e.printStackTrace();
            }
        }
    /**
     * Reconnect to an AMQP broker.in case the connection breaks at some
       point
     *
     * @throws TimeoutException
     */
    private void reconnect() throws TimeoutException {
        log.info("Reconnecting to AMQP broker...");
        try {
            setupAMQP();
        } catch (IOException e) {
            log.warn("Failed to reconnect to AMQP broker", e);
        }
    }
    /**
     * Setup a connection with an AMQP broker.
     *
     * @throws IOException
     *     This is the method where we actually connect to the queue
     *            using AMQP client api's
     * @throws TimeoutException
     */
    private void setupAMQP() throws IOException, TimeoutException {
        final int prefetchCount = this.prefetchCount;
        final ConnectionFactory connectionFactory = new
ConnectionFactory() {
            public void configureSocket(Socket socket) throws
IOException {
                socket.setTcpNoDelay(false);
                socket.setReceiveBufferSize(20 * 1024);
                socket.setSendBufferSize(20 * 1024);
            }
        };
        connectionFactory.setHost(amqpHost);
        connectionFactory.setPort(amqpPort);
```

```java
            connectionFactory.setUsername(amqpUsername);
            connectionFactory.setPassword(amqpPasswd);
            connectionFactory.setVirtualHost(amqpVhost);
             this.amqpConnection = connectionFactory.newConnection();
            this.amqpChannel = amqpConnection.createChannel();
             log.info("Setting basic.qos prefetch-count to " +
                prefetchCount);
            amqpChannel.basicQos(prefetchCount);
                amqpChannel.exchangeDeclare(EXCHANGE_NAME, "direct");
            amqpChannel.queueDeclare(QUEUE_NAME, true, false, false, null);
                amqpChannel.queueBind(QUEUE_NAME, EXCHANGE_NAME, "");
                this.amqpConsumer = new QueueingConsumer(amqpChannel);
            assert this.amqpConsumer != null;
            this.amqpConsumerTag = amqpChannel.basicConsume(QUEUE_NAME,
                    this.autoAck, amqpConsumer);
            System.out.println("***************");
    }
    /*
     * Cancels the queue subscription, and disconnects from the AMQP
        broker. */
    public void close() {
        try {
            if (amqpChannel != null) {
                if (amqpConsumerTag != null) {
                    amqpChannel.basicCancel(amqpConsumerTag);
                }
                amqpChannel.close();
            }
        } catch (IOException e) {
            log.warn("Error closing AMQP channel", e);
        } catch (TimeoutException e) {
            // TODO Auto-generated catch block
            e.printStackTrace();
        }
          try {
            if (amqpConnection != null) {
                amqpConnection.close();
            }
        } catch (IOException e) {
            log.warn("Error closing AMQP connection", e);
        }
    }
    /*
     * Emit message received from queue into collector
```

```java
 */
public void nextTuple() {
    // if (spoutActive &amp;&amp; amqpConsumer != null) {
    try {
            final QueueingConsumer.Delivery delivery = amqpConsumer
                    .nextDelivery(WAIT_FOR_NEXT_MESSAGE);
            if (delivery == null)
                return;
            final long deliveryTag =
            delivery.getEnvelope().getDeliveryTag();
            String message = new String(delivery.getBody());
              if (message != null &amp;&amp; message.length() > 0) {
                collector.emit(new Values(message), deliveryTag);
            } else {
                log.debug("Malformed deserialized message, null or
                zero-length. "
                            + deliveryTag);
                  if (!this.autoAck) {
                    ack(deliveryTag);
                }
            }
    } catch (ShutdownSignalException e) {
        log.warn("AMQP connection dropped, will attempt to
        reconnect...");
        Utils.sleep(WAIT_AFTER_SHUTDOWN_SIGNAL);
        try {
                reconnect();
        } catch (TimeoutException e1) {
                // TODO Auto-generated catch block
                e1.printStackTrace();
         }
    } catch (ConsumerCancelledException e) {
        log.warn("AMQP consumer cancelled, will attempt to
        reconnect...");
        Utils.sleep(WAIT_AFTER_SHUTDOWN_SIGNAL);
        try {
                reconnect();
        } catch (TimeoutException e1) {
                // TODO Auto-generated catch block
                e1.printStackTrace();
          }
    } catch (InterruptedException e) {
        log.error("Interrupted while reading a message, with
        Exception : " + e);
```

```
            }
        // }
    }
    /*
     * ack method to acknowledge the message that is successfully processed
     */
    public void ack(Object msgId) {
        if (msgId instanceof long) {
            final long deliveryTag = (long) msgId;
            if (amqpChannel != null) {
                try {
                    amqpChannel.basicAck(deliveryTag, false);
                } catch (IOException e) {
                    log.warn("Failed to ack delivery-tag " +
                    deliveryTag, e);
                } catch (ShutdownSignalException e) {
                    log.warn(
                            "AMQP connection failed. Failed to
                            ack delivery-tag "
                                    + deliveryTag, e);
                }
            }
        } else {
            log.warn(String.format("don't know how to ack(%s: %s)",
            msgId.getClass().getName(), msgId));
        }
    }
    public void fail(Object msgId) {
        if (msgId instanceof long) {
            final long deliveryTag = (long) msgId;
            if (amqpChannel != null) {
                try {
                    if (amqpChannel.isOpen()) {
                        if (!this.autoAck) {
amqpChannel.basicReject(deliveryTag,
                                    requeueOnFail);
                        }
                    } else {
                        reconnect();
                    }
                } catch (IOException e) {
                    log.warn("Failed to reject delivery-tag " +
                    deliveryTag, e);
```

```
                } catch (TimeoutException e) {
                    // TODO Auto-generated catch block
                    e.printStackTrace();
                }
            }
        } else {
            log.warn(String.format("don't know how to reject(%s:
            %s)", msgId.getClass().getName(), msgId));
        }
    }

    public void declareOutputFields(OutputFieldsDeclarer declarer) {
        declarer.declare(new Fields("messages"));

    }

    public void activate() {
        // TODO Auto-generated method stub

    }

    public void deactivate() {
        // TODO Auto-generated method stub

    }

    public Map<String, Object> getComponentConfiguration() {
        // TODO Auto-generated method stub
        return null;
    }
}
```

接下来将快速浏览前面代码段的关键方法及其内部工作机制。

- `public AMQPspout(...)`：这是一个构造函数，其中由主机 IP、端口、用户名和 RabbitMQ 的密码等对关键变量进行初始化操作。除此之外，它还设置了 Requeue 标识，以防止由于某种原因拓扑无法处理消息。

- `public void open(...)`：这是 IRichSpout 的基本方法。这里的预取计数值告诉我们应该读取多少条记录，并将其保存在 spout 缓冲区中，以供拓扑使用。

- `private void setupAMQP(...)`：这是一个关键方法，通过声明连接工厂、交换器和队列，设置 spout 和 RabbitMQ 连接，并将它们绑定到信道上实现该方法。

- `public void nextTuple()` 这是从 RabbitMQ 信道道上接收消息并将消息发送到收集器中以供拓扑使用的方法。
- 以下代码段检索消息及其正文，并将其发送到拓扑中。

```
..
final long deliveryTag = delivery.getEnvelope().getDeliveryTag();
String message = new String(delivery.getBody());
..
collector.emit(new Values(message), deliveryTag);
```

- 下一步，捕获拓扑生成器，以将 **AMQPSpout** 组件与其他 bolt 绑定在一起。

```
TopologyBuilder builder = new TopologyBuilder();
    builder.setSpout("spout", new AMQPSpout("localhost", 5672,
"guest", "guest", "/", true, false), 1);
```

- 接下来，将连续的传感器数据流插入到 RabbitMQ 队列。建议你连接任何免费的流数据源，如 Facebook 或 Twitter。

以下是一些 `pom.xml` 条目，这些条目是整个程序正确执行 Eclipse 设置所需的依赖项。

以下是 Storm、RabbitMQ、Jackson 和 PubNub 的 Maven 依赖项。

```
...
  <properties>
    <project.build.sourceEncoding>UTF-8</project.build.sourceEncoding>
    <storm.version>0.9.3</storm.version>
</properties>
  <dependencies>
  <dependency>
      <groupId>com.rabbitmq</groupId>
      <artifactId>amqp-client</artifactId>
      <version>3.6.2</version>
</dependency>
<dependency>
      <groupId>org.apache.storm</groupId>
      <artifactId>storm-core</artifactId>
      <version>0.9.3</version>
      <scope>provided</scope>
</dependency>
<dependency>
  <groupId>com.pubnub</groupId>
  <artifactId>pubnub-gson</artifactId>
  <version>4.4.4</version>
```

```xml
</dependency>
<dependency>
        <groupId>com.fasterxml.jackson.core</groupId>
        <artifactId>jackson-databind</artifactId>
        <version>2.6.3</version>
</dependency>
..
```

6.4 PubNub 数据流发布者

本节将构建一个快速发布者，从 PubNub 中读取传感器数据消息流并将其推送到 RabbitMQ。

```java
..
 public class TestStream {

     private static final String EXCHANGE_NAME = "MYExchange";
     private final static String QUEUE_NAME = "MYQueue";
     private final static String ROUTING_KEY = "MYQueue";
private static void RMQPublisher(String myRecord) throws IOException,
TimeoutException
{

     ConnectionFactory factory = new ConnectionFactory();
     Address[] addressArr = { new Address("localhost", 5672) };
     Connection connection = factory.newConnection(addressArr);
     Channel channel = connection.createChannel();
     channel.exchangeDeclare(EXCHANGE_NAME, "direct");
     channel.queueDeclare(QUEUE_NAME, true, false, false, null);
     channel.queueBind(QUEUE_NAME, EXCHANGE_NAME, ROUTING_KEY);
     int i = 0;
     while (i < 1) {
         try {
             channel.basicPublish(EXCHANGE_NAME, ROUTING_KEY,
                     MessageProperties.PERSISTENT_TEXT_PLAIN,
                     myRecord.getBytes());
             System.out.println(" [x] Sent '" + myRecord + "' sent at
               " + new Date());
             i++;
             Thread.sleep(2);
         } catch (Exception e) {
             e.printStackTrace();

         }
```

```java
            }
            channel.close();
            connection.close();
        }
public static void main ( String args[])
{
PNConfiguration pnConfiguration = new PNConfiguration();
pnConfiguration.setSubscribeKey("sub-c-5f1b7c8e-fbee-11e3-
aa40-02ee2ddab7fe");
PubNub pubnub = new PubNub(pnConfiguration);
pubnub.addListener(new SubscribeCallback() {
    @Override
    public void status(PubNub pubnub, PNStatus status) {
        if (status.getCategory() ==
PNStatusCategory.PNUnexpectedDisconnectCategory) {
            // This event happens when radio / connectivity is lost
        }

        else if (status.getCategory() ==
PNStatusCategory.PNConnectedCategory) {
//System.out.println("2");
            // Connect event. You can do stuff like publish, and know
you'll get it.
            // Or just use the connected event to confirm you are
subscribed for
            // UI / internal notifications, etc
            if (status.getCategory() ==
PNStatusCategory.PNConnectedCategory){
System.out.println("status.getCategory()="+status.getCategory());
            }
        }
        else if (status.getCategory() ==
PNStatusCategory.PNReconnectedCategory) {
            // Happens as part of our regular operation. This event happens
when
            // radio / connectivity is lost, then regained.
        }
        else if (status.getCategory() ==
PNStatusCategory.PNDecryptionErrorCategory) {
            // Handle messsage decryption error. Probably client configured
to
            // encrypt messages and on live data feed it received plain
text.
        }
```

```
        }

    @Override
    public void message(PubNub pubnub, PNMessageResult message) {
        // Handle new message stored in message.message
        String strMessage = message.getMessage().toString();
                System.out.println("******"+strMessage);
        try {
                RMQPublisher(strMessage);
            } catch (IOException e) {
                // TODO Auto-generated catch block
                e.printStackTrace();
            } catch (TimeoutException e) {
                // TODO Auto-generated catch block
                e.printStackTrace();
            }

        }
    @Override
    public void presence(PubNub pubnub, PNPresenceEventResult presence) {

        }
});

pubnub.subscribe().channels(Arrays.asList("pubnub-sensornetwork")).
execute();
    }
}
```

这段代码实际上使用订阅键连接到 PubNub 流(),如下所示:

```
STREAM DETAILS
Channel: pubnub-sensor-network
Subscribe key: sub-c-5f1b7c8e-fbee-11e3-aa40-02ee2ddab7fe
```

要在项目中使用此流,请复制和粘贴以前的代码段或订阅此信道和子键。

PubNub 侦听器先前绑定到由订阅键标识的订阅信道,并将消息发送到 RabbitMQ——MyExchange,MYQueue。正在执行的程序截图和 RabbitMQ 截图如图 6-10 和图 6-11 所示,还有一个示例消息作为参考。

第 6 章 集成 Storm 与数据源

图 6-10

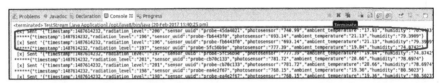

图 6-11

图 6-12 所示为 RabbitMQ 中显示的消息，读者可以在其中看到 MyQueue 中排列的 PubNub 传感器流中的消息。

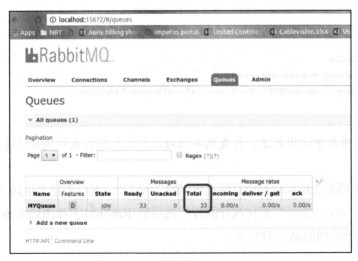

图 6-12

6.5 将 Storm 和 RMQ_PubNub 传感器数据拓扑串在一起

 消息以实时连续流的形式出现——对于本书的示例,实际上在几秒后就关闭了 PubNub 流发布者,因此读者只看到 33 条消息。

读者可以进一步研究队列以检索并查看消息。单击 MYQueue,然后滚动到下一页的 **Get Message** 部分,并在控制台中输入希望检索和查看的消息数量,如图 6-13 所示。

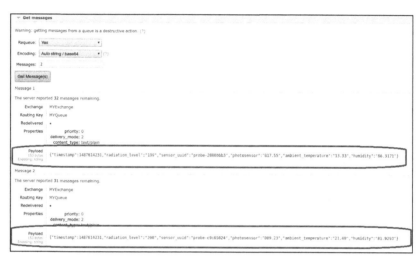

图 6-13

6.5 将 Storm 和 RMQ_PubNub 传感器数据拓扑串在一起

在 JSON Bolt 中解析拓扑中的 JSON 工作负载,并查看执行结果。代码如下:

```
...
public class JsonBolt implements IRichBolt {
..
    private static final long serialVersionUID = 1L;
    OutputCollector collector;

    public void prepare(Map arg0, TopologyContext arg1, OutputCollector arg2) {
        this.collector = arg2;
    }

    public void execute(Tuple arg0) {
```

第 6 章　集成 Storm 与数据源

```
            String jsonInString = arg0.getStringByField("messages");
            System.out.println("message read from queue" + jsonInString);
            JsonConverter jsonconvertor = new JsonConverter();
            MySensorData mysensorObj = jsonconvertor.run(jsonInString);
            this.collector.emit(new Values(mysensorObj));
            this.collector.ack(arg0);
        }
..

}
..

class JsonConverter {
..

    public MySensorData run(String jsonInString) {
        ObjectMapper mapper = new ObjectMapper();
        MySensorData mysensorObj = null;
        try {

            mysensorObj = mapper.readValue(jsonInString,
            MySensorData.class);

            // Pretty print
            String prettyStaff1 =
            mapper.writerWithDefaultPrettyPrinter()
                    .writeValueAsString(mysensorObj);
            System.out.println(prettyStaff1);

        } catch (JsonGenerationException e) {
            ..
        return mysensorObj;
        }
}
```

示例中的 JSONBolt 基本上接收来自拓扑 AMQPSpout 的消息，并将 JSON 字符串转换为可以基于业务逻辑进行进一步处理的 JSON 对象，例如，本示例进一步加强了 SensorProcessorBolt。

```
TopologyBuilder builder = new TopologyBuilder();
builder.setSpout("spout", new AMQPSpout("localhost", 5672, "guest",
"guest", "/", true, false), 1);
builder.setBolt("split", new JsonBolt(), 8).shuffleGrouping("spout");
builder.setBolt("count", new SensorProcessorBolt(),
2).fieldsGrouping("split", new Fields("word"));
```

SensorProcessorBolt 检查传感器中 radiationLevel 发出的数据和过滤器，

6.5 将 Storm 和 RMQ_PubNub 传感器数据拓扑串在一起

并仅发射辐射级别大于 197 的事件。

```java
..
public class SensorProcessorBolt extends BaseBasicBolt {

    public void execute(Tuple tuple, BasicOutputCollector collector) {
        MySensorData mysensordata = (MySensorData) tuple.getValue(0);
        if (mysensordata.getRadiation_level() > 197) {
System.out.println("#####################################");
            System.out.println(mysensordata.getSensor_uuid());
            System.out.println(mysensordata.toString());
System.out.println("#####################################");
        }

        collector.emit(new Values(mysensordata.toString()));
    }
..
}
```

正在执行的完整拓扑如图 6-14 所示。

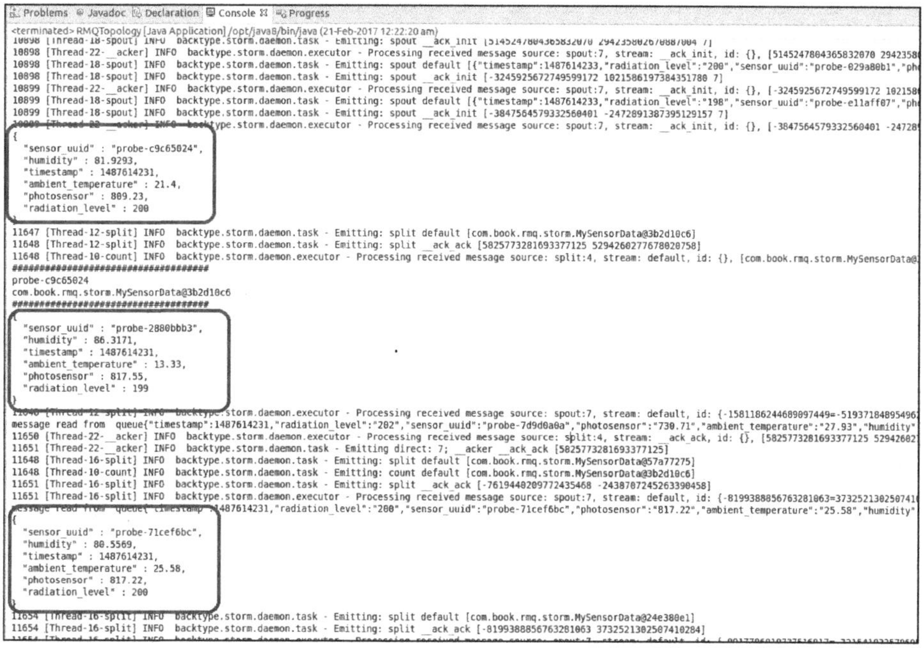

图 6-14

由图 6-13 可以很容易地推断出，我们只发射高（高于 197）辐射级的事件作为此拓扑的输出。读者可以进一步增强此用例，并对物联网传感器数据流、其他各种免费或在 PubNub 上提供的数据源应用各种业务分析，如图 6-15 所示。

图 6-15

6.6 小结

在本章，我们带领读者掌握了 RabbitMQ 消息传输系统——先安装基本的软件，再写一个基本的发布者-订阅者应用程序，然后进一步掌握将代理服务集成到 Storm 拓扑的方法。接着，引入 PubNub 传感器数据，以此作为 Storm 拓扑可以计算和处理的物联网事件连续实时数据流，并进一步探讨端到端的应用程序构建练习。我们力求先引导读者浏览拓扑中每个组件的示例和代码，然后启发并鼓励读者进一步深入研究，使读者通过 PubNub 获取更多的数据源并在已有的拓扑处理框架下挑战更高级的业务逻辑。

第 7 章 从 Storm 到 Sink

在本章,我们将稳定存储和集成到 Storm 拓扑——先安装 Cassandra,然后通过 Storm 进行集成。

本章主要包括以下内容

- 安装并配置 Cassandra
- Storm 和 Cassandra 拓扑
- Storm 和 IMDB 集成处理维度数据
- 集成表示层与 Storm
- 小试牛刀

7.1 安装并配置 Cassandra

在开始安装和配置 Cassandra 之前,首先解释一下 Cassandra 是什么以及为什么它如此受欢迎。Cassandra 是一个列式存储的 NoSQL 数据库。如果用户需要高可用性和高扩展性,那么 Cassandra 将是最佳选择。Cassandra 具有很高的读写性能,但只保证最终一致性。

最终一致性指当用户在数据库中插入一条记录同时另一个用户读取它时,可能会使新添加的记录对用户可见或者不可见。Cassandra 中的 **Keyspace** 与 RDMS 中的数据库相同,其余术语与 RDMS 相同。Cassandra 是一个开源组件,如果用户需要使用设计好的 UI 管理集群,那么请使用 DataStax。**DataStax** 提供付费高级服务,可以全职支持。接下来看看如何设置 Cassandra。

7.1.1 安装 Cassandra

下载最新的 Cassandra 3.10 版本。下载 apache-cassandra-3.10-bin.tar.gz，执行如下命令来解压缩。

```
mv apache-cassandra-3.10-bin.tar.gz ~/demo
tar -xvf ~/demo/apache-cassandra-3.10-bin.tar.gz
```

解压缩后的文件列表，如图 7-1 所示。

图 7-1

在本地单机运行 Cassandra 程序。

```
/bin/Cassandra
```

Cassandra 作为后台进程启动，按 Enter 键退出日志。为了验证 Cassandra 是否正常工作，可以执行如下命令。

```
/bin/nodetool status
```

输出如图 7-2 所示。

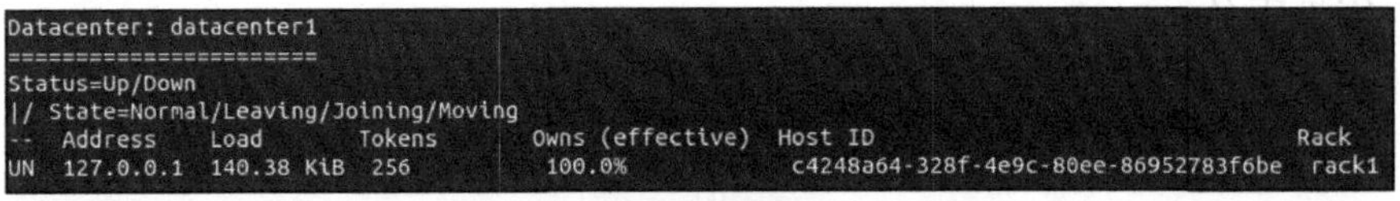

图 7-2

Cassandra 提出了虚拟节点（vnodes）的概念。vnodes 由 Cassandra 在每个节点上自动创建。默认情况下，启动 Cassandra 时会创建 256 个 vnodes。若要检查当前节点上的 vnodes，请使用以下命令：

```
/bin/nodetool ring
```

伴随加载过程，会出现一长串的令牌 ID，如图 7-3 所示。

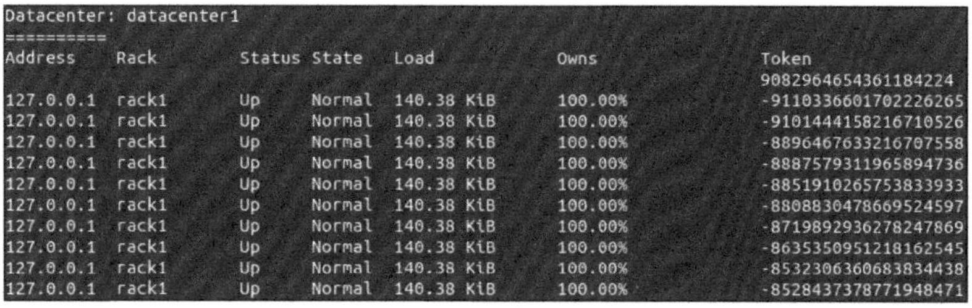

```
Datacenter: datacenter1
==========
Address    Rack     Status State   Load         Owns      Token
                                                          9082964654361184224
127.0.0.1  rack1    Up     Normal  140.38 KiB   100.00%   -9110336601702226265
127.0.0.1  rack1    Up     Normal  140.38 KiB   100.00%   -9101444158216710526
127.0.0.1  rack1    Up     Normal  140.38 KiB   100.00%   -8896467633216707558
127.0.0.1  rack1    Up     Normal  140.38 KiB   100.00%   -8887579311965894736
127.0.0.1  rack1    Up     Normal  140.38 KiB   100.00%   -8851910265753833933
127.0.0.1  rack1    Up     Normal  140.38 KiB   100.00%   -8808830478669524597
127.0.0.1  rack1    Up     Normal  140.38 KiB   100.00%   -8719892936278247869
127.0.0.1  rack1    Up     Normal  140.38 KiB   100.00%   -8635350951218162545
127.0.0.1  rack1    Up     Normal  140.38 KiB   100.00%   -8532306360683834438
127.0.0.1  rack1    Up     Normal  140.38 KiB   100.00%   -8528437378771948471
```

图 7-3

7.1.2 配置 Cassandra

安装目录中的 Conf 文件夹包含了所有需要的配置文件。将配置划分为若干部分以更好地理解。以下几行是不同配置属性的解释。

- cluster_name：集群的名称。
- seeds：以逗号分隔的集群种子的 IP 地址列表。这些节点协助 gossip 协议检查节点健康状况。

> 2 个或 3 个节点足以完成此任务。此外，在整个集群中使用相同的种子节点。

- listen_address：节点的 IP 地址。这是允许其他节点与此节点通信的原因，所以它的更改非常关键。
- 切勿将 listen_address 设置为 0.0.0.0。
- 用户可以设置 listen_interface 来告诉 Cassandra 使用哪个接口，以及连续使用哪个地址。
- 设置 listen_address 或 listen_interface，但不能同时设置。
- native_transport_port：对于 storage_port，客户端使用它与 Cassandra 进行通信。
- 确保此端口未被防火墙禁用掉。
- 数据日志文件的位置。data_file_directories：保存数据文件的一个或多个目录的位置；commitlog_directory：保存 commitlog 文件的目录的位置；saved_caches_directory：保存已缓存的目录的位置；hints_directory：

第 7 章 从 Storm 到 Sink

保存提示的目录位置。

 出于性能原因，如果有多个磁盘，请考虑将 commitlog 和数据文件放在不同的磁盘上。

- **环境变量**：要设置环境变量，请使用位于 installation_dir/bin/cassandra.in.sh 中的 cassondra.in.sh，主要用于设置 JVM 级别的环境变量。
- **日志采集**：Logback 框架用作 Cassandra 中的日志采集器。用户可以通过在 installation_dir/conf/logback.xml 中的 logback.xml 来更改日志级别设置。

7.2 Storm 和 Cassandra 拓扑

如第 4 章所述，Storm 具有 spout 和 bolt，必须使用 Cassandra bolt 才能保存记录。将 Storm 和 Cassandra 集成在一起有两种常见的方法：第一种方法是使用 storm-cassandra 内置库，只需调用 CassandraBolt 和所需的参数即可；第二种方法是使用 DataStax Cassandra 库，需要使用构建管理器导入该库，并使用包装器类与 Cassandra 建立连接。以下是使用 DataStax 库集成 Storm 和 Cassandra 的步骤。

① 添加以下依赖项。代码如下：

```
<dependency>
    <groupId>com.datastax.cassandra</groupId>
    <artifactId>cassandra-driver-core</artifactId>
    <version>3.1.0</version>
</dependency>
<dependency>
    <groupId>com.datastax.cassandra</groupId>
    <artifactId>cassandra-driver-mapping</artifactId>
    <version>3.1.0</version>
</dependency>
<dependency>
    <groupId>com.datastax.cassandra</groupId>
    <artifactId>cassandra-driver-extras</artifactId>
    <version>3.1.0</version>
</dependency>
```

7.2 Storm 和 Cassandra 拓扑

在 pom.xml 中添加依赖项。

② 创建类。创建一个类并使用 BaseBasicBolt 对其进行扩展。类 CassandraBolt 要求用户实现 execute 和 declareOutputFields 这两个方法，代码如下：

```
public class CassandraBolt extends BaseBasicBolt
```

③ 实现 prepare 和 cleanup 方法。代码如下：

```
public void prepare( Map stormConf, TopologyContext context )
{
    cluster =
Cluster.builder().addContactPoint( "127.0.0.1" ).build();         // Line #1
    session = cluster.connect( "demo" );                          // Line #2
}
@Override
public void cleanup()
{
    cluster.close();                                              // Line #3
}
```

覆盖 prepare 和 cleanup 这两个方法。第 1 行是创建 Cassandra 集群，必须在其中提供所有 Cassandra 节点的 IP。如果配置的端口号不是 9042，则为 IP 提供端口号，该端口号由以下内容分隔：第 2 行是从集群中创建一个会话，它将在其中创建一个与 Cassandra 节点集群列表中提供的一个节点的连接。此外，还需要在创建会话时提供 keyspace 名称。第 3 行在作业完成后关闭群集。在 Storm 中，Prepare 方法仅在拓扑部署在集群上时调用一次，cleanup 方法仅在终止拓扑时调用一次。

④ execute 方法的定义。execute 方法对每个要处理到 Storm 的元组执行：

```
session.execute( "INSERT INTO users (lastname, age, city, email,
firstname) VALUES ('Jones', 35, 'Austin', 'bob@example.com',
'Bob')" );
```

前面的代码将一行语句插入到 demo 密钥空间中的 Cassandra 表 user 中。字段在元组参数中可用。因此，用户可以读取字段并更改前面的语句，如下所示：

```
String userDetail = (String) input.getValueByField("event");
String[] userDetailFields = userDetail.split(":");
session.execute("INSERT INTO users (lastname, age, city, email,
firstname) VALUES ('userDetailFields[0]', userDetailFields[1], '
userDetailFields[2]', ' userDetailFields[3]', '
userDetailFields[4]')");
```

可以在 `session.execute` 方法中执行任何有效的 SQL。会话还提供了 `PreparedStatement`，以用于大容量插入/更新/删除。

7.3 Storm 和 IMDB 集成处理维度数据

IMDB 代表**基于内存类型的数据库**。IMDB 需要在处理事件流的同时保持中间结果，或者保持与事件相关的静态信息（事件中可能没有提供这些信息）。员工详细信息可以根据员工 ID 和进出办公室的事件存储在 IMDB 中。在这种情况下，事件不包含员工的完整信息以节省网络成本和提高性能，因此，当 Storm 处理事件时，它将从 IMDB 中获取有关员工的静态信息，并将其与事件的详细信息一起保存在 Cassandra 或任何其他数据库中，以进行进一步的分析。市场上有许多开源的 IMDB 工具，其中比较著名的有 **Hazelcast**、**Memcached** 和 **Redis**。

让我们来看看如何整合 Storm 和 Hazelcast。不需要对 Hazelcast 进行特殊设置。执行以下步骤。

① 添加依赖项。

```xml
<dependency>
    <groupId>com.hazelcast</groupId>
    <artifactId>hazelcast</artifactId>
    <version>3.8</version>
</dependency>
<dependency>
    <groupId>com.hazelcast</groupId>
    <artifactId>hazelcast-client</artifactId>
    <version>3.8</version>
</dependency>
```

在 `pom.xml` 中添加依赖项以将 `hazelcast` 包含在项目中。

② 创建 Hazelcast 集群并加载数据。

```
Config cfg = new Config(); // Line #1
HazelcastInstance instance = Hazelcast.newHazelcastInstance(cfg);
// Line #2
Map<Integer, String> mapCustomers = instance.getMap("employees");
// Line #3
mapCustomers.put(1, "Joe"); // Line #4
```

```
mapCustomers.put(2, "Ali"); // Line #5
mapCustomers.put(3, "Avi"); // Line #6
```

第 1 行创建了配置对象 `cfg`。这只是一个空对象，但是可以提供已在远程机器上运行 Hazelcast 集群的 IP 和端口。第 2 行使用配置创建 Hazelcast 实例。使用其默认配置，然后搜索在本地主机端口 5701 上运行的 Hazelcast。如果任何实例已在 5701 上运行，则生成端口号为 5702 的集群；否则，将创建在本地主机端口 5701 上运行的单个实例。

第 3 行正在创建一个名为 `employees` 的 Hazelcast 映射。键为 `Integer`，值为 `String`。

第 4 行、第 5 行和第 6 行说明了如何在 `Map` 中插入/添加值，它与 Java 中的 `Map` 相同。

③ 在 Storm 拓扑中使用 Hazelcast `Map`。

```
public class EmployeeBolt extends BaseBasicBolt {
    public void prepare( Map stormConf, TopologyContext context )
    {
        ClientConfig clientConfig = new ClientConfig();          // Line #1
clientConfig.getNetworkConfig().addAddress( "127.0.0.1:5701" );
        // Line #2
        HazelcastInstance client =
    HazelcastClient.newHazelcastClient( clientConfig );
    }                                                             // Line #3
    public void execute( Tuple input, BasicOutputCollector arg1 )
    {
        String inoutDetails = (String)
input.getValueByField( "event" );
        IMap map = client.getMap( "employees" );                  // Line #4
        System.out.println( "Employee Name:" +
        map.get( inoutDetails.split( ":" )[0] ) );   // Line #5
    }
    @Override
    public void cleanup()     {
        cluster.close();                                          // Line #7
    }
}
```

首先创建一个类，并使用 `BaseBasicBolt` 对其进行扩展，就像我们在前面所做的那样。第 1 行创建 `ClientConfig` 对象，以创建 `HazelcastInstance`。第 2 行在 `clientConfig` 对象中添加主机名和 Hazelcast 服务器的 IP。创建 `HazelcastClientInstance` 端实例以与 Hazelcast 服务器通信。使用第 4 行中的客户端对象获取 `employees` 的 `Map` 引用。现在，读者可以以 JavaMap 相同的方式使用 Map 来执行添加/更新/删除操作。读者可以使用市面上提供的 `storm-redis` 库集成 Storm 和 Redis。

到此读者已经掌握了如何在 Cassandra 中使用 Storm 持久化数据，以及如何使用 Hazelcast 来获取有关事件的静态信息。接下来，我们继续将集成 Storm 与表示层。

7.4 集成表示层与 Storm

将数据可视化是了解数据最好的方式，而且可以依此做出重要决策。市场上有许多可视化工具，每个可视化工具都需要一个数据库来存储和处理数据。有些组合是在 Elasticsearch 上使用 Grafana、有些是 Elasticsearch 上使用 Kibana，有些是在 Influxdb 上使用 Grafana。在本章，我们将讨论 Grafana、Elasticsearch 和 Storm 的融合。

本例将使用来自 PubNub 的数据流，该数据流提供实时传感器数据。PubNub 提供了从通道读取数据的所有类型的 API。在这里，需要一个程序从 PubNub 订阅通道获取数据并将其推入 Kafka 主题。你可以在配套资源中找到程序。

使用 Elasticsearch 插件设置 Grafana

Grafana 是一个分析平台，可以理解用户数据并将其可视化到仪表板上。

1. 下载 Grafana

下载 Grafana。安装独立的二进制文件。

```
wget https://s3-us-west-2.amazonaws.com/grafana-releases/release/grafana-4.2.0.linux-x64.tar.gz
    cp grafana-4.2.0.linux-x64.tar.gz ~/demo/.
    tar -zxvf grafana-4.2.0.linux-x64.tar.gz
```

解压缩文件，如图 7-4 所示。

```
:~/demo/grafana-4.2.0$ ls
bin  conf  data  LICENSE.md  NOTICE.md  public  README.md  scripts  vendor
```

图 7-4

2. 配置 Grafana

需要在 `defaults.ini` 配置文件的路径部分下更改以下两种配置。

- `data`：Grafana 存储 `sqlite3` 数据库（如果使用）、基于文件的会话（如果使用）和其他数据的路径。
- `logs`：Grafana 将存储日志的路径。

3. 在 Grafana 中安装 Elasticsearch 插件

使用以下命令在 Grafana 中安装最新的 Elasticsearch 版本插件：

```
/bin/grafana-cli plugins install stagemonitor-elasticsearch-app
```

4. 运行 Grafana

使用以下命令运行 Grafana：

```
/bin/grafana-server
```

这需要一段时间才能启动，因为它需要建立自己的数据库。一旦成功启动，就可以使用地址 `http://localhost:3000` 访问 UI。输入 `admin` 作为用户名，`admin` 作为密码。

控制面板如图 7-5 所示。

图 7-5

5. 在 Grafana 中添加 Elasticsearch 数据源

现在，单击仪表板上的 **Add data source**，添加/更新该值，如图 7-6 所示。

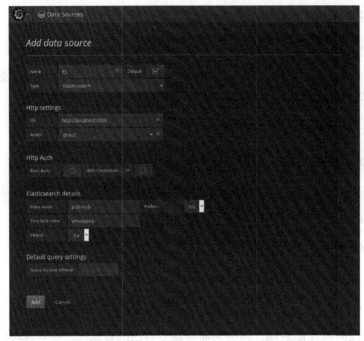

图 7-6

- 将数据源的 **Name** 输入为 ES。
- 从 **Type** 下拉列表中选择 **Elasticsearch**。
- 在 **URL** 中输入 http://localhost:9200。
- 选择 **Access** 为 **direct**。

  为此，读者必须在 elasticsearch.yml 中添加以下行：

```
http.cors.enabled: true
http.cors.allow-origin: "*"
```

- **Index name** 输入为 pub-nub。
- 在 **Time field name** 中输入 timestamp。
- 选择 **Version** 为 **5.x**。
- 单击 **Add** 按钮。

之后，用户可以通过单击 **save and test** 来测试连接。

6. 写代码

Elasticsearch 的设置和配置在前面已有描述。除了上面提到的配置，我们还可以在 `elasticsearch.yml` 中更改给定的属性。

```
cluster.name: my-application
```

在前面的章节中，读者学习了如何编写拓扑代码。现在将讨论 Elasticsearch bolt，它从其他 bolt/spout 中读取数据并写入 Elasticsearch。

① 在 `pom.xml` 中添加以下依赖项。

```xml
<!-- Need for JSON parsing -->
<dependency>
    <groupId>com.fasterxml.jackson.core</groupId>
    <artifactId>jackson-databind</artifactId>
    <version>2.8.7</version>
</dependency>
<!-- Need for Elasticsearch java client API -->
<dependency>
    <groupId>org.elasticsearch.client</groupId>
    <artifactId>transport</artifactId>
    <version>5.2.2</version>
</dependency>

<!-- Need for Store core -->
<dependency>
    <groupId>org.apache.storm</groupId>
    <artifactId>storm-core</artifactId>
    <version>1.0.3</version>
    <exclusions>
        <exclusion>
            <groupId>org.slf4j</groupId>
            <artifactId>log4j-over-slf4j</artifactId>
        </exclusion>
    </exclusions>
</dependency>
<dependency>
    <groupId>org.apache.logging.log4j</groupId>
    <artifactId>log4j-api</artifactId>
    <version>2.7</version>
</dependency>
<dependency>
    <groupId>org.apache.logging.log4j</groupId>
    <artifactId>log4j-core</artifactId>
```

```xml
    <version>2.7</version>
</dependency>

<!-- Need for Store Kafka integration -->
<dependency>
    <groupId>org.apache.storm</groupId>
    <artifactId>storm-kafka</artifactId>
    <version>1.0.3</version>
</dependency>
<dependency>
    <groupId>org.apache.kafka</groupId>
    <artifactId>kafka_2.11</artifactId>
    <version>0.8.2.2</version>
    <exclusions>
        <exclusion>
            <artifactId>jmxri</artifactId>
            <groupId>com.sun.jmx</groupId>
        </exclusion>
        <exclusion>
            <artifactId>jms</artifactId>
            <groupId>javax.jms</groupId>
        </exclusion>
        <exclusion>
            <artifactId>jmxtools</artifactId>
            <groupId>com.sun.jdmk</groupId>
        </exclusion>
        <exclusion>
            <groupId>org.slf4j</groupId>
            <artifactId>slf4j-log4j12</artifactId>
        </exclusion>
    </exclusions>
</dependency>
```

② 定义类和类级别变量。

```
public class ElasticSearchBolt extends BaseBasicBolt
```

③ 声明类级别变量。

```
Client client;
PreBuiltTransportClient preBuiltTransportClient;
ObjectMapper mapper;
```

④ 利用 BaseBasicBolt 实现 prepare 方法。

```
@Override
```

```
public void prepare( @SuppressWarnings("rawtypes") Map stormConf,
TopologyContext context ) {
    mapper = new ObjectMapper();     // line #1
    Settings settings = Settings.builder().put( "cluster.name",
"my-application" ).build();   // line #2
    preBuiltTransportClient = new
PreBuiltTransportClient( settings );  // line #3
    client = preBuiltTransportClient.addTransportAddress( new
    InetSocketTransportAddress( new
    InetSocketAddress( "localhost", 9300 ) ) );    // line #4
}
```

第 1 行为 JSON 解析创建 `ObjectMapper` 对象。任何其他库都可用于解析。第 2 行创建设置的对象,这是客户端执行 Elasticsearch 所必需的。这些设置要求集群名称为必填项。第 3 行创建 Elasticsearch 客户端对象,该对象将设置作为参数。在第 4 行,需要用户提供 Elasticsearch 节点的详细信息,其中包括主机名和端口号。如果存在一个 Elasticsearch 集群,则使用 `InetAddress`。

⑤ 利用 `BaseBasicBolt` 实现 `execute` 方法。

```
@Override
public void execute(Tuple input, BasicOutputCollector collector) {
    String valueByField = input.getString(0); // Line #1
    try {
        IndexResponse response = client.prepareIndex("pub-nub", "sensordata").
        setSource(convertStringtoMap(valueByField)).get(); // Line #2
    }
    catch (IOException e) {
        e.printStackTrace();
    }
}
```

第 1 行是从包含事件的零位置的元组中获取值。第 2 行是创建索引(如果不存在的话),以及将文档添加到索引中。

`client.preareIndex` 正在创建一个索引,该索引要求索引名称为第一个参数,类型为第二个参数。`setSource` 是在索引中添加文档的方法。`get` 方法将返回 `IndexResponse` 类的对象,该类包含创建索引和添加文档的请求是否已成功完成的信息。`convertstringtomap` 将字符串转换为更改字段数据类型的映射,需要使其在 `Grafana` 仪表板上可见。如果事件已采用所需的格式,则不需要转换类型。

现在必须集成 Elasticsearch bolt 和 Kafka spout 来读取数据。

⑥ 编写绑定 Kafka spout 和 Elasticsearch bolt 的拓扑。

```
TopologyBuilder topologyBuilder = new TopologyBuilder(); //Line #1
BrokerHosts hosts = new ZkHosts("localhost:2181"); //Line #2
SpoutConfig spoutConfig = new SpoutConfig(hosts, "sensor-data" ,
"/" + topicName, UUID.randomUUID().toString()); //Line #3
spoutConfig.scheme = new SchemeAsMultiScheme(new StringScheme()); //Line #4
KafkaSpout kafkaSpout = new KafkaSpout(spoutConfig); //Line #5
topologyBuilder.setSpout("spout", kafkaSpout, 1); //Line #6
topologyBuilder.setBolt("es-bolt", new ElasticSearchBolt(), 1).
shuffleGrouping("spout");//Line #7
```

第 1 行是创建 `topologyBuilder` 对象，其中包含 spout 和所有 bolt 的信息。本例使用了 Storm 提供的预定义 Storm-Kafka 集成 spout。第 2 行将 Zookeeper 主机详细信息创建为 `BrokerHosts`。在第 3 行，创建了 Kafka spout 所需的 `spoutConfig`，其中包含有关 ZooKeeper 主机、主题名称、ZooKeeper 根目录和客户端 ID 的信息，以便与 Kafka 代理进行通信。第 4 行将方案设置为字符串，默认情况下为字节。使用 `spoutConfig` 作为参数创建 `kafkaSpout` 对象。现在，首先将 spout 设置为 `topologyBuilder`，参数为 ID、`kafkaSpout` 和第 6 行中的并行性提示。同样，在 `topologyBuilder` 中设置 Elasticsearch bolt，如第 7 行所述。

⑦ 提交拓扑。第 1 行创建 Storm 所需的 `config` 对象。第 2 行设置 Storm 监管者引擎中所需的最低工作单元数量。现在，创建 `localCluster` 作为提交拓扑的集群。使用参数作为拓扑、配置和 `topologyBuilder` 名称调用 `submitTopology` 方法。读者将在配套资源中找到完整的代码。

```
Config config = new Config();                                   // Line #1
config.setNumWorkers( 3 );                                      // Line #2
LocalCluster cluster = new LocalCluster();                      // Line #3
cluster.submitTopology( "storm-es-example", config,
        topologyBuilder.createTopology() );                     // Line #4
```

7. 执行代码

导入 Eclipse 中配套资源提供的代码。在 Eclipse 中导入程序完毕后，执行以下步骤。

- 运行 `PubNubDataStream.java`。
- 运行 `SenorTopology.java`。

8. 在 Grafana 上可视化输出

配置控制面板，如图 7-7 所示。

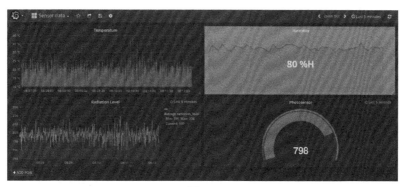

图 7-7

代码如下:

```java
package com.book.chapter7.visualization.example;

import java.util.Arrays;
import java.util.Properties;

import org.apache.kafka.clients.producer.KafkaProducer;
import org.apache.kafka.clients.producer.ProducerRecord;

import com.pubnub.api.PNConfiguration;
import com.pubnub.api.PubNub;
import com.pubnub.api.callbacks.SubscribeCallback;
import com.pubnub.api.enums.PNStatusCategory;
import com.pubnub.api.models.consumer.PNStatus;
import com.pubnub.api.models.consumer.pubsub.PNMessageResult;
import com.pubnub.api.models.consumer.pubsub.PNPresenceEventResult;
```

定义 PubNub 数据流。

```java
public class PubNubDataStream {

    public KafkaProducer<Integer, String> producer;

    public PubNubDataStream()   {
        Properties properties = new Properties();
        properties.put( "bootstrap.servers", "localhost:9092" );
        properties.put( "key.serializer",
          "org.apache.kafka.common.serialization.StringSerializer" );
        properties.put( "value.serializer",
```

第 7 章 从 Storm 到 Sink

```
            "org.apache.kafka.common.serialization.StringSerializer" );
        properties.put( "acks", "1" );
        producer = new KafkaProducer<Integer, String>( properties );
    }

    private void pubishMessageToKafka( String message )  {
        ProducerRecord<Integer, String> data = new ProducerRecord<Integer,
        String>(
            "sensor-data", message );
            producer.send( data );
    }

    private void getMessageFromPubNub()  {
        PNConfiguration pnConfiguration = new PNConfiguration();
        pnConfiguration
        .setSubscribeKey( "sub-c-5f1b7c8e-fbee-11e3-aa40-02ee2ddab7fe" );
        PubNub pubnub = new PubNub( pnConfiguration );
pubnub.addListener( new SubscribeCallback()  {
    @Override
    public void status( PubNub pubnub, PNStatus status )  {
        System.out.println( pubnub.toString() + "::" + status.toString() );

        if ( status.getCategory() ==
        PNStatusCategory.PNUnexpectedDisconnectCategory )  {
            // This event happens when radio / connectivity is lost
        }
        else if ( status.getCategory() ==
        PNStatusCategory.PNConnectedCategory )  {
            // Connect event. You can do stuff like publish, and know
            // you'll get it.
            // Or just use the connected event to confirm you are
            // subscribed for
            // UI / internal notifications, etc
            if ( status.getCategory() ==
            PNStatusCategory.PNConnectedCategory )  {
              System.out.println( "status.getCategory()=" +
              status.getCategory() );
            }
        } else if ( status.getCategory() ==
            PNStatusCategory.PNReconnectedCategory )  {
            // Happens as part of our regular operation. This event
            // happens when
            // radio / connectivity is lost, then regained.
        } else if ( status.getCategory() ==
            PNStatusCategory.PNDecryptionErrorCategory )  {
            // Handle messsage decryption error. Probably client
```

```
                // configured to
                // encrypt messages and on live data feed it received plain
                // text.
            }
        }
        @Override
        public void message( PubNub pubnub, PNMessageResult message ) {
            // Handle new message stored in message.message
            String strMessage = message.getMessage().toString();
            System.out.println( "******" + strMessage );
            pubishMessageToKafka( strMessage );
            /*
             * log the following items with your favorite logger -
             * message.getMessage() - message.getSubscription() -
             * message.getTimetoken()
             */
        }

        @Override
        public void presence( PubNub pubnub, PNPresenceEventResult presence ) {

        }
    } );

    pubnub.subscribe().channels( Arrays.asList( "pubnub-sensornetwork" ) ).
    execute();
    }
    public static void main( String[] args ) {
        new PubNubDataStream().getMessageFromPubNub();
    }
}
```

Elasticsearch 持久化 bolt。

```
package com.book.chapter7.visualization.example;

import java.io.IOException;
import java.net.InetSocketAddress;
import java.util.Date;
import java.util.HashMap;
import java.util.Map;

import org.apache.storm.task.TopologyContext;
import org.apache.storm.topology.BasicOutputCollector;
import org.apache.storm.topology.OutputFieldsDeclarer;
import org.apache.storm.topology.base.BaseBasicBolt;
import org.apache.storm.tuple.Tuple;
```

```java
import org.elasticsearch.action.index.IndexResponse;
import org.elasticsearch.client.Client;
import org.elasticsearch.common.settings.Settings;
import org.elasticsearch.common.transport.InetSocketTransportAddress;
import org.elasticsearch.transport.client.PreBuiltTransportClient;

import com.fasterxml.jackson.core.JsonParseException;
import com.fasterxml.jackson.core.type.TypeReference;
import com.fasterxml.jackson.databind.JsonMappingException;
import com.fasterxml.jackson.databind.ObjectMapper;

public class ElasticSearchBolt extends BaseBasicBolt {
    private static final long serialVersionUID = -9123903091990273369L;
    Client client;
    PreBuiltTransportClient preBuiltTransportClient;
    ObjectMapper mapper;

    @Override
    public void prepare( @SuppressWarnings("rawtypes") Map stormConf,
        TopologyContext context ) {
        // instance a json mapper
        mapper = new ObjectMapper(); // create once, reuse
        Settings settings = Settings.builder()
        .put( "cluster.name", "my-application" ).build();
        preBuiltTransportClient = new
        PreBuiltTransportClient( settings );
        client = preBuiltTransportClient.addTransportAddress( new
InetSocketTransportAddress( new InetSocketAddress( "localhost", 9300 ) ) );
}


@Override
public void cleanup() {
    preBuiltTransportClient.close();
    client.close();
}


@Override
public void declareOutputFields( OutputFieldsDeclarer declarer ) {
}


@Override
public void execute( Tuple input, BasicOutputCollector collector ) {
    String valueByField = input.getString( 0 );
    System.out.println( valueByField );
    try {
        IndexResponse response = client.prepareIndex( "pub-nub", "sensordata" )
        .setSource( convertStringtoMap( valueByField ) ).get();
```

```
        System.out.println( response.status() );
    } catch ( IOException e ) {
        e.printStackTrace();
    }
}

public Map<String, Object> convertStringtoMap( String fieldValue ) throws
JsonParseException, JsonMappingException, IOException  {
    System.out.println( "Orignal value " + fieldValue );
    Map<String, Object> convertedValue = new HashMap<>();
    Map<String, Object> readValue = mapper.readValue( fieldValue, new
    TypeReference<Map<String, Object> >() { } );

    convertedValue.put( "ambient_temperature",
Double.parseDouble( String.valueOf( readValue.get( "ambient_temperature" ) ) ) );
convertedValue.put( "photosensor",
Double.parseDouble( String.valueOf( readValue.get( "photosensor" ) ) ) );
convertedValue.put( "humidity",
Double.parseDouble( String.valueOf( readValue.get( "humidity" ) ) ) );
convertedValue.put( "radiation_level",
Integer.parseInt( String.valueOf( readValue.get( "radiation_level" ) ) ) );
convertedValue.put( "sensor_uuid", readValue.get( "sensor_uuid" ) );
convertedValue.put( "timestamp", new Date() );

System.out.println( "Converted value " + convertedValue );
return(convertedValue);
    }
}
```

拓扑构建器绑定 spout 和 bolt。

```
package com.book.chapter7.visualization.example;

import java.util.UUID;

import org.apache.storm.Config;
import org.apache.storm.LocalCluster;
import org.apache.storm.kafka.BrokerHosts;
import org.apache.storm.kafka.KafkaSpout;
import org.apache.storm.kafka.SpoutConfig;
import org.apache.storm.kafka.StringScheme;
import org.apache.storm.kafka.ZkHosts;
import org.apache.storm.spout.SchemeAsMultiScheme;
import org.apache.storm.topology.TopologyBuilder;
```

```java
public class SensorTopology {
    public static void main( String args[] ) throws InterruptedException  {
    Config config = new Config();
    config.setNumWorkers( 3 );
    TopologyBuilder topologyBuilder = new TopologyBuilder();

    String zkConnString = "localhost:2181";
    String topicName = "sensor-data";

    BrokerHosts hosts = new ZkHosts( zkConnString );
    SpoutConfig spoutConfig = new SpoutConfig( hosts, topicName, "/" +
    topicName, UUID.randomUUID().toString() );
    spoutConfig.scheme = new SchemeAsMultiScheme( new StringScheme() );
    KafkaSpout kafkaSpout = new KafkaSpout( spoutConfig );
    topologyBuilder.setSpout( "spout", kafkaSpout, 1 );
    topologyBuilder.setBolt( "es-bolt", new ElasticSearchBolt(),
    1 ).shuffleGrouping( "spout" );

        LocalCluster cluster = new LocalCluster();
        cluster.submitTopology( "storm-es-example", config,
        topologyBuilder.createTopology() );
    }
}
```

7.5 小试牛刀

在本节，我们将结合 Storm、Kafka、Hazelcast 和 Cassandra 构建一个实际的用例。该用例使用具有唯一标识的电话号码，将通信实时分组数据输入至 Kafka。系统必须将每个电话号码的总使用情况（字节）存储到 Hazelcast 中，并将总使用情况持久化到 Cassandra 中，此外还将每个事件持久化到 Cassandra 中。

伪代码如下。

- 创建将数据保存在 Cassandra 中的 `CassandraBolt`。

- 创建一个 bolt，根据电话号码读取 Hazelcast 中的值，并将其与当前值相加，同时在 Hazelcast 中更新相同的条目。

- 创建一个拓扑，将 Kafka spout 连接到上一步中提到的自定义 bolt 中，然后创建 `CassandraBolt` 以持久化总使用情况。还将 Kafka spout 连接到 `CassandraBolt`，以持久化每一个事件。

7.5 小试牛刀

导入代码。

```
package com.book.chapter7.diy;

Here we have the import files

import java.util.Date;
import java.util.Properties;
import java.util.concurrent.ThreadLocalRandom;

import org.apache.kafka.clients.producer.KafkaProducer;
import org.apache.kafka.clients.producer.ProducerRecord;
```

生成数据的代码。

```
public class DataGenerator {
    public static void main( String args[] ) {
    Properties properties = new Properties();
    properties.put( "bootstrap.servers", "localhost:9092" );
    properties.put( "key.serializer",
    "org.apache.kafka.common.serialization.StringSerializer" );
    properties.put( "value.serializer",
    "org.apache.kafka.common.serialization.StringSerializer" );
    properties.put( "acks", "1" );

    KafkaProducer<Integer, String> producer = new KafkaProducer<Integer,
    String>( properties );
    int counter = 0;
    int nbrOfEventsRequired = Integer.parseInt( args[0] );
    while ( counter < nbrOfEventsRequired )  {
    StringBuffer stream = new StringBuffer();

    long phoneNumber = ThreadLocalRandom.current().nextLong( 99999999501,
    99999999991 );
    int bin = ThreadLocalRandom.current().nextInt( 1000, 9999 );
    int bout = ThreadLocalRandom.current().nextInt( 1000, 9999 );

    stream.append( phoneNumber );
    stream.append( "," );
    stream.append( bin );
    stream.append( "," );
    stream.append( bout );
    stream.append( "," );
    stream.append( new Date( ThreadLocalRandom.current().nextLong() ) );
```

```
        System.out.println( stream.toString() );
        ProducerRecord<Integer, String> data = new ProducerRecord<Integer,
        String>(
            "storm-diy", stream.toString() );
            producer.send( data );
            counter++;
    }

    producer.close();
    }
}
```

使用以下代码片段启动 Hazelcast 服务器。

```
package com.book.chapter7.diy;

import com.hazelcast.core.Hazelcast;

public class HCServer {
    public static void main( String args[] )  {
        Hazelcast.newHazelcastInstance();
    }
}
```

bolt 之间的数据传输对象。

```
Sending data from the import java.io.Serializable

package com.book.chapter7.diy;

import java.io.Serializable;

public class PacketDetailDTO implements Serializable {

    private static final long serialVersionUID = 9148607866335518739L;
    private long phoneNumber;
    private int bin;
    private int bout;
    private int totalBytes;
    private String timestamp;

    public long getPhoneNumber()  {
        return(phoneNumber);
    }

    public void setPhoneNumber( long phoneNumber )  {
```

```java
        this.phoneNumber = phoneNumber;
    }

    public int getBin() {
        return(bin);
    }

    public void setBin( int bin ) {
        this.bin = bin;
    }

    public int getBout() {
        return(bout);
    }

    public void setBout( int bout ) {
        this.bout = bout;
    }

    public int getTotalBytes() {
        return(totalBytes);
    }

    public void setTotalBytes( int totalBytes ) {
        this.totalBytes = totalBytes;
    }

    public String getTimestamp() {
        return(timestamp);
    }

    public void setTimestamp( String timestamp ) {
        this.timestamp = timestamp;
    }
}
```

分析器和使用计算 bolt

在代码中，有类 map。

```java
package com.book.chapter7.diy;

import java.util.Map;

import org.apache.storm.task.TopologyContext;
```

```java
import org.apache.storm.topology.BasicOutputCollector;
import org.apache.storm.topology.OutputFieldsDeclarer;
import org.apache.storm.topology.base.BaseBasicBolt;
import org.apache.storm.tuple.Fields;
import org.apache.storm.tuple.Tuple;
import org.apache.storm.tuple.Values;

import com.hazelcast.client.HazelcastClient;
import com.hazelcast.client.config.ClientConfig;
import com.hazelcast.core.HazelcastInstance;
import com.hazelcast.core.IMap;

public class ParseAndUsageBolt extends BaseBasicBolt {

    private static final long serialVersionUID = 1271439619204966337L;
    HazelcastInstance client;
    IMap<String, PacketDetailDTO> usageMap;

@Override
public void prepare( Map stormConf, TopologyContext context ) {
    ClientConfig clientConfig = new ClientConfig();
    clientConfig.getNetworkConfig().addAddress( "127.0.0.1:5701" );
    client = HazelcastClient.newHazelcastClient( clientConfig );
    usageMap = client.getMap( "usage" );
}

@Override
public void execute( Tuple input, BasicOutputCollector collector ) {
    PacketDetailDTO packetDetailDTO = new PacketDetailDTO();
    String valueByField = input.getString( 0 );
    String[] split = valueByField.split( "," );
    long phoneNumber = Long.parseLong( split[0] );
    PacketDetailDTO packetDetailDTOFromMap = usageMap.get( phoneNumber );
    if ( null == packetDetailDTOFromMap ) {
        packetDetailDTOFromMap = new PacketDetailDTO();
    }
    packetDetailDTO.setPhoneNumber( phoneNumber );
    int bin = Integer.parseInt( split[1] );
    packetDetailDTO.setBin( (packetDetailDTOFromMap.getBin() + bin) );
    int bout = Integer.parseInt( split[2] );
    packetDetailDTO.setBout( packetDetailDTOFromMap.getBout() + bout );
    packetDetailDTO.setTotalBytes( packetDetailDTOFromMap.getTotalBytes()
    + bin + bout );
```

```
        usageMap.put( split[0], packetDetailDTO );

        PacketDetailDTO tdrPacketDetailDTO = new PacketDetailDTO();
        tdrPacketDetailDTO.setPhoneNumber( phoneNumber );
        tdrPacketDetailDTO.setBin( bin );
        tdrPacketDetailDTO.setBout( bout );
        tdrPacketDetailDTO.setTimestamp( split[3] );

        collector.emit( "usagestream", new Values( packetDetailDTO ) );
        collector.emit( "tdrstream", new Values( tdrPacketDetailDTO ) );
    }

    @Override
    public void cleanup()  {
        client.shutdown();
    }

    @Override
    public void declareOutputFields( OutputFieldsDeclarer declarer ) {
        declarer.declareStream( "usagestream", new Fields( "usagestream" ) );
        declarer.declareStream( "tdrstream", new Fields( "tdrstream" ) );
    }
}
```

Cassandra 持久化 bolt。

```
package com.book.chapter7.diy;

import java.util.Map;

import org.apache.storm.task.TopologyContext;
import org.apache.storm.topology.BasicOutputCollector;
import org.apache.storm.topology.OutputFieldsDeclarer;
import org.apache.storm.topology.base.BaseBasicBolt;
import org.apache.storm.tuple.Tuple;

import com.datastax.driver.core.Cluster;
import com.datastax.driver.core.Session;

public class TDRCassandraBolt extends BaseBasicBolt {
    private static final long  serialVersionUID = 1L;
    private Cluster cluster;
    private Session session;
    private String hostname;
    private String keyspace;
```

```java
    public TDRCassandraBolt( String hostname, String keyspace ) {
        this.hostname = hostname;
        this.keyspace = keyspace;
    }

    @Override
    public void prepare( Map stormConf, TopologyContext context ) {
        cluster = Cluster.builder().addContactPoint( hostname ).build();
        session = cluster.connect( keyspace );
    }

    public void execute( Tuple input, BasicOutputCollector arg1 ) {
        PacketDetailDTO packetDetailDTO = (PacketDetailDTO)
        input.getValueByField( "tdrstream" );
        session.execute( "INSERT INTO packet_tdr (phone_number, bin, bout,
        timestamp) VALUES ("
        + packetDetailDTO.getPhoneNumber()
        + ", "
        + packetDetailDTO.getBin()
        + ","
        + packetDetailDTO.getBout()
        + ",'" + packetDetailDTO.getTimestamp() + "')" );
    }

    public void declareOutputFields( OutputFieldsDeclarer arg0 ) {
    }

    @Override
    public void cleanup() {
        session.close();
        cluster.close();
    }
}
```

`java.util.Map` 的代码：使用 Cassandra 中的持久化。

```java
package com.book.chapter7.diy;

import java.util.Map;

import org.apache.storm.task.TopologyContext;
import org.apache.storm.topology.BasicOutputCollector;
import org.apache.storm.topology.OutputFieldsDeclarer;
```

```java
import org.apache.storm.topology.base.BaseBasicBolt;
import org.apache.storm.tuple.Tuple;

import com.datastax.driver.core.Cluster;
import com.datastax.driver.core.Session;

public class UsageCassandraBolt extends BaseBasicBolt {
    private static final long serialVersionUID = 1L;
    private Cluster cluster;
    private Session session;
    private String hostname;
    private String keyspace;

    public UsageCassandraBolt( String hostname, String keyspace ) {
        this.hostname = hostname;
        this.keyspace = keyspace;
    }

    @Override
    public void prepare( Map stormConf, TopologyContext context ) {
        cluster = Cluster.builder().addContactPoint( hostname ).build();
        session = cluster.connect( keyspace );
    }

    public void execute( Tuple input, BasicOutputCollector arg1 ) {
        PacketDetailDTO packetDetailDTO = (PacketDetailDTO)
        input.getValueByField( "usagestream" );
        session.execute( "INSERT INTO packet_usage (phone_number, bin, bout,
        total_bytes) VALUES ("
        + packetDetailDTO.getPhoneNumber()
        + ", "
        + packetDetailDTO.getBin()
        + ", "
        + packetDetailDTO.getBout()
        + ", " + packetDetailDTO.getTotalBytes() + ")" );
    }

    public void declareOutputFields( OutputFieldsDeclarer arg0 ) {
    }

    @Override
    public void cleanup() {
        session.close();
        cluster.close();
```

        }
}

拓扑构建器绑定 spout 和 bolts。

```java
package com.book.chapter7.diy;

import java.util.UUID;

import org.apache.storm.Config;
import org.apache.storm.LocalCluster;
import org.apache.storm.kafka.BrokerHosts;
import org.apache.storm.kafka.KafkaSpout;
import org.apache.storm.kafka.SpoutConfig;
import org.apache.storm.kafka.StringScheme;
import org.apache.storm.kafka.ZkHosts;
import org.apache.storm.spout.SchemeAsMultiScheme;
import org.apache.storm.topology.TopologyBuilder;

public class TelecomProcessorTopology {
    public static void main( String[] args )  {
        Config config = new Config();
        config.setNumWorkers( 3 );
        TopologyBuilder topologyBuilder = new TopologyBuilder();

        String zkConnString = "localhost:2181";
        String topicName = "storm-diy";

        BrokerHosts hosts = new ZkHosts( zkConnString );
        SpoutConfig spoutConfig  = new SpoutConfig( hosts, topicName, "/" + topicName, UUID.randomUUID().toString() );
        spoutConfig.scheme = new SchemeAsMultiScheme( new StringScheme() );
        KafkaSpout kafkaSpout = new KafkaSpout( spoutConfig );
        topologyBuilder.setSpout( "spout", kafkaSpout, 1 );
        topologyBuilder.setBolt( "parser", new ParseAndUsageBolt(),
            1 ).shuffleGrouping( "spout" );
        topologyBuilder.setBolt( "usageCassandra", new
            UsageCassandraBolt( "localhost", "usage" ),
            1 ).shuffleGrouping( "parser", "usagestream" );
        topologyBuilder.setBolt( "tdrCassandra", new
            TDRCassandraBolt( "localhost", "tdr" ), 1 ).shuffleGrouping( "parser",
            "tdrstream" );

        LocalCluster cluster = new LocalCluster();
            cluster.submitTopology( "storm-diy", config,
```

```
        topologyBuilder.createTopology() );
    }
}
```

7.6 小结

在本章，我们介绍了 Storm 中所有可能的接收器。Storm 和接收器之间支持内置集成，但这种方法目前还不够成熟，无法在所需的配置上运行，因此本章用普通 Java 代码和任何外部工具进行连接。我们先介绍了 Storm 与最新版 Cassandra 的集成，然后查看了与 Storm 相关的所有类型用例所需的内存数据库，接着解释了 Storm 和 Hazelcast 的集成。之后，我们展示了 Storm 中一个更重要的集成，即 Storm 与表示层的集成。在这里，我们选择了 Elasticsearch 与 Grafana，并完成了这个用例。最后，我们向读者提出了一个问题，引导读者用本章开头介绍的接收器来思考和编写代码。

第 8 章 Storm Trident

在本章，我们将介绍用于微批的 DRPC 和 Storm Trident 抽象，并基于此实现一些实际用例。

本章主要包括以下内容

- 状态保持和 Trident
- 基本 Storm Trident 拓扑
- Trident 内部实现
- Trident 操作
- DRPC
- 小试牛刀

8.1 状态保持和 Trident

维持其状态 Trident 是一个分布式实时分析框架。Trident 以容错的方式在内部（如在内存中）或在外部（如 Hazelcast）。它类似于处理"精确一次"（exactly once）类型事件。Trident 适用于微批，如聚合、过滤等。

这里举一个例子来解释如何实现"精确一次"语义：假设读者正在计算访问博客的人数，并将计数值存储在数据库中。现在假设在数据库中存储了一个表示计数的值，并且每次处理新元组时都会增加计数值。

如果发生故障，则将通过 Storm 拓扑重放元组。问题是元组是否已被处理并且数据是否已经在数据库中更新了。如果是，就不应该再次更新它；如果元组没有成功处理，就必须在数据库中更新计数；如果元组已被处理但是在更新数据库中的计数值时失败，就应该更新数据库。

为了实现确保元组仅在系统中处理一次"精确一次"语义，spout 应该向 bolt/spout 提供信息。在容错方面有 3 种类型的 spout：事务性 spout、非事务性 spout 和不透明事务性 spout。现在，让我们来看看其中两种 spout。

8.1.1 事务性 spout

让我们来看看 Trident spout 如何处理元组以及其特征是什么。

- Trident 以小批量方式处理元组。
- 每个批次都有唯一的事务 ID。
- Trident 确保每个元组都是经过批处理的，因此不会跳过任何元组。
- 具有给定事务 ID 的批次始终相同。如果一个批次被重新处理，那么这个批次将具有相同的事务 ID 和元组集。
- 元组不能是多个批次的一部分。每个批处理都有一个唯一的元组集。

在代码中可以使用 Kafka 定义事务 spout，代码如下：

```
TransactionalTridentKafkaSpout tr = new TransactionalTridentKafkaSpout(new
TridentKafkaConfig(new ZkHosts("localhost:9091"), "test"));
```

`TransactionalTridentKafkaSpout` 在 `storm-core` 库中可用，它提供了所有先前给出的属性。如果节点不可用，spout 无法获得相同的元组集，则事务性 spout 不具有容错能力。为了克服这个问题，存在不透明事务性 spout。

8.1.2 不透明事务性 spout

不透明事务性 spout 具有这样的属性：每个元组只需在一个批次中成功处理。但是，元组可能无法在一个批次中处理，可能在稍后的批次中成功处理。

`OpaqueTridentKafkaSpout` 是一个具有该属性的 spout，并且具有 Kafka 节点丢失的容错能力。每当 `OpaqueTridentKafkaSpout` 发出批处理时，它就从最后一个批处理完成发射的地方开始发出元组。这可确保多个批处理不会跳过或成功地处理任何元组。

可以用 Kafka 定义一个不透明事务性 spout，代码如下：

```
OpaqueTridentKafkaSpout otks = new OpaqueTridentKafkaSpout(new
TridentKafkaConfig(new ZkHosts("localhost:9091"), "test"));
```

8.2 基本 Storm Trident 拓扑

在本节，我们将单词计数作为示例学习基本的 Storm Trident 拓扑。更多示例参见本章后面的内容。示例代码如下：

```
FixedBatchSpout spout = new FixedBatchSpout(new Fields("sentence"), 3,
new Values("this is simple example of trident topology"),
new Values("this example count same words"));
spout.setCycle(true); // Line 1
TridentTopology topology = new TridentTopology(); // Line 2
MemoryMapState.Factory stateFactory = new MemoryMapState.Factory(); // Line 3
topology.newStream("spout1", spout) // Line 4
.each(new Fields("sentence"), new Split(), new Fields("word")) // Line 5
.groupBy(new Fields("word")) // Line 6
.persistentAggregate(stateFactory, new Count(), new
Fields("count")).newValuesStream() // Line 7
.filter(new DisplayOutputFilter()) // Line 8
.parallelismHint(6); // Line 9
Config config = new Config(); // Line 10
config.setNumWorkers(3); // Line 11
LocalCluster cluster = new LocalCluster(); // Line 12
cluster.submitTopology("storm-trident-example", config, topology.build()); // Line 13
```

第 1 行中的 spout 启动程序 `FixedBatchSpout` 用于配套资源测试。如果 `setCycle` 设置为 `True`，则可以给出一组重复值。定义 `TransactionalTridentKafkaSpout`，因为需要连接 Zookeeper 的详细信息和主题名称。另一个构造函数具有相同的参数以及客户端 ID：

```
TransactionalTridentKafkaSpout spout = new
TransactionalTridentKafkaSpout(new TridentKafkaConfig(new
ZkHosts("localhost:9091"), "test"));
```

第 2 行创建了 `TridentTopology` 对象，该对象需要在集群上提交。

第 3 行创建了 `MemoryMapState.Factory` 对象，将数据保存在内存中并保持状态。

在拓扑中设置 spout，其名称为第 4 行中的 `spout1`。

现在，对第 5 行中的每个元组执行一项操作。这里需要对每个元组执行 `split` 函数。

每种拓扑方法都有 3 个参数，即 spout 中的输入字段名称、每个元组要执行的函数以及输出字段的名称。split 方法的实现如下。句子根据空间进行分割。

```
class Split extends BaseFunction {
  public void execute(TridentTuple tuple, TridentCollector collector) {
    String sentence = tuple.getString(0);
    for (String word : sentence.split(" ")) {
      collector.emit(new Values(word));
    }
  }
}
```

 我们将在第 8 章详细讨论函数、过滤器和聚合操作。

第 6 行对名称为 word 的元组执行分组操作，使用相同的单词对所有元组进行分组并创建批处理，产生带有单词和计数的元组。

persistentAggregate 根据第 7 行中的给定函数执行聚合操作。此外，它还在内存中保持状态。它先从内存中读取数值，然后与当前值相加。之后，它会更新内存缓存以保持更新。这里调用了函数计数，以将相同单词的计数相加。计数的实现代码如下：

```
public class Count implements CombinerAggregator<long> {
  @Override
  public long init(TridentTuple tuple) {
    return 1L;
  }
  @Override
  public long combine(long val1, long val2) {
    return val1 + val2;
  }
  @Override
  public long zero() {
    return 0L;
  }
}
```

要显示输出，请实现自定义过滤器以在控制台上打印元组值。第 8 行应用了过滤器。自定义过滤器 DisplayOutputFilter 的实现代码如下：

```
public class DisplayOutputFilter implements Filter {
    @Override
    public void prepare(Map conf, TridentOperationContext context) {
```

```
    }
    @Override
    public void cleanup() {
    }
    @Override
    public Boolean isKeep(TridentTuple tuple) {
        System.out.println(tuple.get(0)+":"+tuple.get(1));
        return true;
    }
}
```

第 9 行设置并行。第 10～13 行创建配置对象并在本地集群上提交先前创建的拓扑。

完整代码在名为 `BasicTridentTopology.java` 的代码包中提供。

输出结果如图 8-1 所示。

```
example:2
is:1
simple:1
count:1
of:1
topology:1
trident:1
same:1
this:2
words:1
example:4
is:2
simple:2
of:2
topology:2
count:2
words:2
trident:2
same:2
this:4
```

图 8-1

8.3 Trident 内部实现

每个 Trident 流都是 Storm 流。执行器和工作单元的概念与 Storm 中的完全相同。Trident 拓扑只不过是一个 Storm bolt。Trident 的操作（如 spout、each 和聚会）实际上是在 Storm bolt 中实现的。

Trident 将拓扑转换为数据流（有向无环）图，以将操作分配给 bolt，然后将这些 bolt 分配给工作单元。它足够智能，可以优化该任务：将操作组合成 bolt，以便尽可能地通过简单的方法传递元组，并在工作单元之间安排 bolt，以尽可能地将元组传递给本地执行器。

Trident 拓扑的 spout 实际上被称为**主批处理协调器**（**MBC**）。它所做的只是发出一个元组，将自身描述为批处理 1，然后是另一个元组将自身描述为批处理 2，以此类推。决定何时发射这些批处理、重试等是令人非常兴奋的，但是 Storm 并不完全知道这些。这些批处理元组转到拓扑的 spout 协调器。spout 协调器了解外部来源记录的位置和排列，并确保每个源记录唯一属于成功的 Trident 批处理。

8.4　Trident 操作

Trident 操作由 Storm bolt 实现。Trident 可以提供多样化的操作，执行复杂的操作并与内存中的缓存聚合。以下是 Trident 可用的操作。

8.4.1　函数

以下是函数的特征。

- 类必须继承自 `BaseFunction`。
- 这是属于本地操作的分区，意味着不涉及网络传输，并且独立地应用于每个批处理分区。
- 需要一组输入，同时发出零或更多输出。
- 在输出中，它发出包含原始输入元组的输出元组。

示例如下：

```
class PerformDiffFunction extends BaseFunction {
  @Override
  public void execute(TridentTuple tuple, TridentCollector collector) {
     int number1 = tuple.getInteger(0);
     int number2 = tuple.getInteger(1); if(number2>number1){
        collector.emit(new Values(number2-number1));
     }
  }
}
```

输入：

[1,2]
[3,4]
[7,3]

输出：

[1,2,1]
[3,4,1]

8.4.2　Map 函数 and FlatMap 函数

map 函数的特征为：它表示元组的一对一转换；类必须实现 `MapFunction`。

示例如下：

```
class UpperCase implements MapFunction {
  @Override
  public Values execute(TridentTuple input) {
    return new Values(input.getString(0).toUpperCase());
  }
}
```

输入：

```
[this is a simple example of trident topology]
```

输出：

```
[THIS IS A SIMPLE EXAMPLE OF TRIDENT TOPOLOGY]
```

FlatMap 函数的特征为：它表示元组的一对多转换；它将生成的元素展平为新的流；类必须实现 `FlatMapFunction`。

示例如下：

```
class SplitMapFunction implements FlatMapFunction {
  @Override
  public Iterable<Values> execute(TridentTuple input) {
    List<Values> valuesList = new ArrayList<>();
    for (String word : input.getString(0).split(" ")) {
      valuesList.add(new Values(word));
    }
    return valuesList;
  }
}
```

输入：

```
[this is s simple example of trident topology]
```

输出：

```
[this]
[is]
[simple]
[example]
[of]
[trident]
[topology]
[this]
[example]
[count]
[same]
[words]
```

8.4.3 peek 函数

peek 函数用于调试操作之间的元组。以下是使用 flatMap 函数的示例。

```
topology.newStream("spout1", spout).flatMap(new SplitMapFunction())
.map(new UpperCase()).peek(
  new Consumer() {
    @Override
    public void accept(TridentTuple tuple) {
        System.out.print("[");
        for (int index = 0; index < tuple.size(); index++) {
            System.out.print(tuple.get(index));
            if (index < (tuple.size() - 1))
            System.out.print(",");
        }
        System.out.println("]");
    }
} );
```

8.4.4 过滤器

过滤器的特征是：它将元组作为输入，并决定是否保留该元组。示例如下：

```
class MyFilter extends BaseFilter {
  public Boolean isKeep(TridentTuple tuple) {
```

```
        return tuple.getInteger(0) == 1 &amp; &amp; tuple.getInteger(1) == 2;
    }
}
```

输入：

```
[1,2]
[3,4]
[7,3]
```

输出：

```
[1,2]
```

8.4.5 窗口操作

在窗口中，Trident 元组在同一窗口中被处理并发送到下一个操作。窗口操作有如下两种类型。

1．翻滚窗口

每次处理的窗口具有固定间隔或计数。一个元组仅在一个窗口中处理，如图 8-2 所示。

```
| e1 e2 | e3 e4 e5 e6 | e7 e8 e9 |...
0       5             10          15    -> time
   w1         w2            w3
```

图 8-2

其中，e1、e2、e3 是事件，0、5、10 和 15 是 5s 的窗口，w1、w2 和 w3 是窗口。在这里，每个事件都只是一个窗口的一部分。

2．滑动窗口

带有间隔和处理后的窗口，根据时间间隔滑动窗口。一个元组在多个窗口中被处理，如图 8-3 所示。

```
........| e1 e2 | e3 e4 e5 e6 | e7 e8 e9 |...
-5       0       5             10          15    -> time
|<------- w1 -->|
         |<--------- w2 ----->|
                  |<-------------- w3 ---->|
```

图 8-3

8.4 Trident 操作

可以看出，这里的窗口是重叠的，一个事件可以是多个窗口的一部分。在 window 函数的示例中，将集成 Kafka feed 并检查控制台上的输出。

以下是用于理解 Storm Trident 与 Kafka 链接的代码。

```
TridentKafkaConfig config = new TridentKafkaConfig(new
ZkHosts("localhost:2181"), "test"); // line 1
config.scheme = new SchemeAsMultiScheme(new StringScheme()); // line 2
config.startOffsetTime = kafka.api.OffsetRequest.LatestTime();// line 3
TransactionalTridentKafkaSpout spout = new
TransactionalTridentKafkaSpout(config); // line 4
```

第 1 行首先了创建 `TridentKafkaConfig` 对象，该对象使用 Zookeeper 主机名和端口以及主题名称来构造。第 2 行将输入的方案设置为字符串类型。第 3 行将 `startOffsetTime` 置为主题最近的事件而不是主题自启动以后的所有事件。第 4 行用先前定义好的配置文件创建 Trident Kafka spout。

以下代码可以帮助读者理解在 Storm Trident 窗口中操作 API。

```
topology.newStream("spout1", spout) // line 1
.each(new Fields("str"),new Split(), new Fields("word")) // line 2
.window(windowConfig, windowStore, new Fields("word"), new
CountAsAggregator(), new Fields("count")) // line 3
```

首先，在第 1 行定义 spout。这里使用的是上一步创建好的 Kafka spout。第 2 行中，将输入字符串通过空格进行拆分，并将输出定义为 spout 中每个事件的 word 字段。以下是常用的窗口操作 API，以支持任意窗口函数。

```
public Stream window(WindowConfig windowConfig, WindowsStoreFactory
windowStoreFactory, Fields inputFields, Aggregator aggregator, Fields
functionFields)
```

`windowConfig` 可以是以下任何一种。

- `SlidingCountWindow.of(int windowCount, int slidingCount)`
- `SlidingDurationWindow.of(BaseWindowedBolt.Duration windowDuration, BaseWindowedBolt.Duration slidingDuration)`
- `TumblingCountWindow.of(int windowLength)`
- `TumblingDurationWindow.of(BaseWindowedBolt.DurationwindowLength)`

windowStore 可以是以下任何一种。它需要处理元组和聚合值。

- HBaseWindowStoreFactory
- InMemoryWindowsStoreFactory

示例代码如下：

```
WindowsStoreFactory mapState = new InMemoryWindowsStoreFactory();
```

在第 3 行，除了有 windowConfig 和 windowStore 之外，输入为 word 字段，输出为 count 字段，CountAsAggregator 聚合函数用来计算窗口中接收的元组数量。

可以根据窗口配置文件辅助理解输出。

- **滑动计数窗口**：在大小为 100 的窗口中执行 10 次滑动计数操作。

```
SlidingCountWindow.of(100, 10)
```

Kafka 控制台上的输入如图 8-4 所示。

输出如图 8-5 所示。

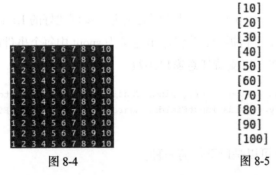

图 8-4　　　　　　　　图 8-5

- **翻滚计数窗口**：在窗口计数 100 次后执行操作。代码如下：

```
TumblingCountWindow.of(100)
```

输出如下：

```
[100]
```

- **滑动持续时间窗口**：如下面代码的所示，窗口持续时间为 6s，滑动持续时间为 3s。

```
SlidingDurationWindow.of(new BaseWindowedBolt.Duration(6,
TimeUnit.SECONDS), new BaseWindowedBolt.Duration(3,
TimeUnit.SECONDS))
```

输出如图 8-6 所示。输出不能像以前那样一致。输出的一致性取决于输入数据的速度。

- **翻滚持续时间窗口**：如下代码所示的窗口持续时间为 3s。

```
TumblingDurationWindow.of(new BaseWindowedBolt.Duration(3,
TimeUnit.SECONDS))
```

输出如图 8-7 所示。

```
[20]
[60]
[50]
[50]              [20]
[51]              [20]
[32]              [20]
[61]              [10]
[50]              [20]
[32]              [10]
[22]              [20]
```

图 8-6　　　　　　　图 8-7

输出不能像以前那样一致。输出的一致性取决于输入数据的速度。

8.4.6　聚合操作

聚合（aggregation）是建立在批处理、分区或者流上的一种操作。Trident 有 3 种不同类型的聚合操作。

1．单独聚合

每个批处理元组都是单独聚合（aggregate）的。在聚合过程中，最初使用全局分组重新分区元组。我们将在后面详细讨论分组。

```
stream.aggregate(new Count(), new Fields("count"))
```

前面的语句将给出批处理的计数。

2．分区聚合

分区聚合（partition aggregate）在每个批处理元组的每个分区上运行一个函数。分区聚合的输出包含单个字段元组。示例如下：

```
stream.partitionAggregate(new Fields("b"), new Sum(), new Fields("sum"))
```

输入：

```
Partition 0:
["a", 1]
["b", 2]

Partition 1:
["a", 3]
["c", 8]

Partition 2:
["e", 1]
["d", 9]
["d", 10]
```

输出：

```
Partition 0:
[3]

Partition 1:
[11]

Partition 2:
[20]
```

3．持久聚合

持久聚合（persistence aggregate）聚合流中所有批处理的所有元组，并将结果存储在内存或数据库中。在本书中，基本的 Trident 拓扑中有一个曾使用内存存储来执行计数的例子。有以下 3 种不同的接口可以执行该聚合器。

（1）**组合器聚合器**（combiner aggregator） 以下是组合器聚合器的特征。

- 返回一个字段作为输出。

- 通过网络传输元组之前执行的部分聚合。

- 为所有元组执行 init 函数。

- 为所有元组执行 comine 函数，除非元组只有单个值。

- 包中提供接口。

```
public interface CombinerAggregator<T> extends Serializable {
  T init(TridentTuple tuple);
  T combine(T val1, T val2);
  T zero();
}
```

示例如下:

```
public class Count implements CombinerAggregator<long> {
  public long init(TridentTuple tuple) {
    return 1L;
  }
  public long combine(long val1, long val2) {
    return val1 + val2;
  }
  public long zero() {
    return 0L;
  }
}
```

（2）**归纳器聚合器**（reducer aggregator） 以下是**归纳器聚合器**的特征。

- 产生单个值。

- 只执行一次 init 函数来获取初始值。

- 在所有元组上执行 reduce 函数。

- 通过网络传输所有元组，然后执行 reduce 函数。

- 与组合器聚合器相比更少被优化。

- 包中提供接口。

```
public interface ReducerAggregator<T> extends Serializable {
  T init();
  T reduce(T curr, TridentTuple tuple);
}
```

示例如下:

```
public class Count implements ReducerAggregator<long> {
  public long init() {
    return 0L;
  }
  public long reduce(long curr, TridentTuple tuple) {
```

```
        return curr + 1;
    }
}
```

4. 聚合器（aggregator）

以下是聚合器的特征：会发出任意数量的元组，且该元组具有任意数量的字段；在批处理之前执行 init 方法；为批处理分区中的每个输入元组执行聚合方法；此方法可以更新状态并可选地发出元组；当聚合方法处理了批处理分区中的所有元组时，会执行完整的方法；包中提供接口。

```
public interface Aggregator<T> extends Operation {
    T init(Object batchId, TridentCollector collector);
    void aggregate(T state, TridentTuple tuple, TridentCollector collector);
    void complete(T state, TridentCollector collector);
}
```

示例如下：

```
public class CountAgg extends BaseAggregator<CountState> {
  static class CountState {
    long count = 0;
  }
  public CountState init(Object batchId, TridentCollector collector) {
    return new CountState();
  }
  public void aggregate(CountState state, TridentTuple tuple,
TridentCollector collector) {
    state.count+=1;
  }
  public void complete(CountState state, TridentCollector collector) {
    collector.emit(new Values(state.count));
  }
}
```

8.4.7 分组操作

分组操作是 Storm Trident 的内置操作，由 `groupBy` 函数执行，使用 `partitionBy` 重新分区元组，然后在分区中对所有具有相同分组字段的元组进行分组。代码示例如下：

```
topology.newStream("spout", spout)
  .each(new Fields("sentence"), new Split(), new Fields("word")) .groupBy(new Fields("word"))
  .persistentAggregate(stateFactory, new Count(), new Fields("count"));
```

根据 Storm Trident 文档，可使用图 8-8 所示的样式按功能能进行分组。

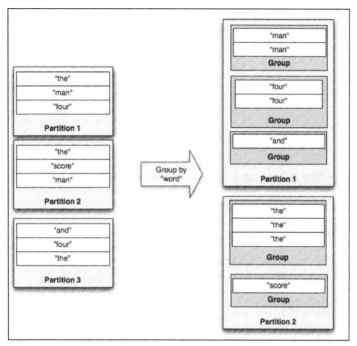

图 8-8

8.4.8 合并和组合操作

合并（merge）操作用于将多个流合并在一起。代码示例如下：

```
topology.merge(stream1, stream2, stream3);
```

组合流的另一种方法是使用组合（join）。例如，stream1 有字段（key, val1, val2），stream2 有（x, val1）。现在使用 stream1 和 stream2 执行组合操作，如下所示：

```
topology.join(stream1, new Fields("key"), stream2, new Fields("x"), new Fields
("key", "a", "b", "c"));
```

stream1 和 stream2 分别基于 key 和 x 进行组合。该字段的输出被定义为来自 stream1 的键，val1 和 val2 为来自 stream1 中的 a 和 b 的键，val1 为来自 stream2 的 c。

输入：

```
Stream 1:
[1, 2, 3]
```

```
Stream 2:
[1, 4]
```

输出：

```
[1, 2, 3, 4]
```

8.5　DRPC

分布式远程生成调用（DRPC）用 Storm 执行计算密集型的函数。它将函数名称和相应的参数作为输入，输出是每个函数调用的结果。

DRPC 客户端的代码示例如下：

```
DRPCClient client = new DRPCClient("drpc.server.location", 3772);
System.out.println(client.execute("words", "cat dog the man"));
```

DRPC Storm 的代码如下：

```
topology.newDRPCStream("words")
.each(new Fields("args"), new Split(), new Fields("word"))
.groupBy(new Fields("word"))
.stateQuery(wordCounts, new Fields("word"), new MapGet(), new Fields("count"))
.aggregate(new Fields("count"), new Sum(), new Fields("sum"));
```

代码的执行流程如图 8-9 所示。

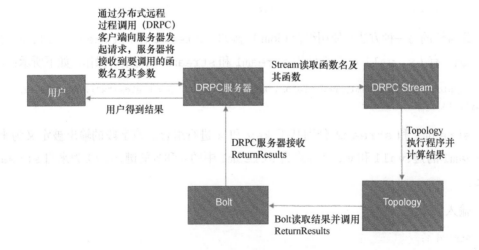

图 8-9

8.6 小试牛刀

在本节，我们用过滤器、分组和聚合器构建一个用例。用例找到批处理中生成最大数据的前 10 个设备。

- 编写一个数据生成器，以发布包含电话号码、输入字节和输出字节等字段的事件。
- 数据生成器将在 Kafka 中发布事件。
- 编写拓扑程序。其流程为：从 Kafka 中获取事件；应用过滤器将排名前 10 位的电话号码排除；基于逗号对事件进行分割；对电话号码执行分组操作，即将相同的电话号码聚合在一起；执行聚合操作，即对输入和输出字节进行求和；然后应用 FirstN 函数进行组装，该函数需要字段名称和元素个数以进行计算；最后在控制台上显示它。

详细代码读者可以参考如下配套资源：

```
package com.book.chapter8.diy;
```

如下代码片段展示了需要导入的资源。

```
import org.apache.storm.Config;
import org.apache.storm.LocalCluster;
import org.apache.storm.kafka.StringScheme;
import org.apache.storm.kafka.ZkHosts;
import org.apache.storm.kafka.trident.TransactionalTridentKafkaSpout;
import org.apache.storm.kafka.trident.TridentKafkaConfig;
import org.apache.storm.spout.SchemeAsMultiScheme;
import org.apache.storm.trident.TridentTopology;
import org.apache.storm.trident.operation.BaseFilter;
import org.apache.storm.trident.operation.BaseFunction;
import org.apache.storm.trident.operation.TridentCollector;
import org.apache.storm.trident.operation.builtin.Debug;
import org.apache.storm.trident.operation.builtin.FirstN;
import org.apache.storm.trident.operation.builtin.Sum;
import org.apache.storm.trident.tuple.TridentTuple;
import org.apache.storm.tuple.Fields;
import org.apache.storm.tuple.Values;
```

用以绑定 spout 和 bolt 的拓扑构建器

基础类如下所示：

```java
public class TridentDIY {
    public static void main(String args[]) {
        TridentKafkaConfig config = new TridentKafkaConfig(new ZkHosts(
            "localhost:2181"), "storm-trident-diy");
        config.scheme = new SchemeAsMultiScheme(new StringScheme());
        config.startOffsetTime = kafka.api.OffsetRequest.LatestTime();
        TransactionalTridentKafkaSpout spout = new
TransactionalTridentKafkaSpout(config);
        TridentTopology topology = new TridentTopology();
        topology.newStream("spout", spout).filter(new ExcludePhoneNumber())
                        .each(new Fields("str"), new DeviceInfoExtractor(),
new Fields("phone", "bytes"))
                        .groupBy(new Fields("phone"))
                        .aggregate(new Fields("bytes", "phone"), new Sum(),
new Fields("sum")).applyAssembly(new FirstN(10, "sum"))
.each(new Fields("phone", "sum"), new Debug());

Config config1 = new Config();
        config1.setNumWorkers(3);
        LocalCluster cluster = new LocalCluster();
        cluster.submitTopology("storm-trident-diy", config1,
topology.build());
    }
}
```

从事件流中过滤电话号码

格式化数据的类。

```java
class ExcludePhoneNumber extends BaseFilter {
    private static final long serialVersionUID = 7961541061613235361L;

    public Boolean isKeep(TridentTuple tuple) {
        return !tuple.get(0).toString().contains("9999999950");
    }
}
```

从事件流中提取设备信息

如下代码呈现了数据编码的方式。

```java
class DeviceInfoExtractor extends BaseFunction {
    private static final long serialVersionUID = 4889855511293326495L;

    @Override
    public void execute(TridentTuple tuple, TridentCollector collector) {
        String event = tuple.getString(0);
        System.out.println(event);
        String[] splittedEvent = event.split(",");
        if(splittedEvent.length>1){
            long phoneNumber = long.parselong(splittedEvent[0]);
            int bin = Integer.parseint(splittedEvent[1]);
            int bout = Integer.parseint(splittedEvent[2]);
            int totalBytesTransferred = bin + bout;
            System.out.println(phoneNumber+":"+bin+":"+bout);
            collector.emit(new Values(phoneNumber, totalBytesTransferred));
        }
    }
}
```

```
package com.book.chapter8.diy;
```

导入文件。

```java
import java.util.Date;
import java.util.Properties;
import java.util.concurrent.ThreadLocalRandom;

import org.apache.kafka.clients.producer.KafkaProducer;
import org.apache.kafka.clients.producer.ProducerRecord;
```

数据生成的类如下：

```java
public class DataGenerator {
    public static void main(String args[]) {
        Properties properties = new Properties();
        properties.put("bootstrap.servers", "localhost:9092");
        properties.put("key.serializer",
"org.apache.kafka.common.serialization.StringSerializer");
        properties.put("value.serializer",
"org.apache.kafka.common.serialization.StringSerializer");
        properties.put("acks", "1");
        KafkaProducer<Integer, String> producer = new
KafkaProducer<Integer, String>(properties);
        int counter =0;
        int nbrOfEventsRequired = Integer.parseint(args[0]);
```

```
            while (counter<nbrOfEventsRequired) {
                StringBuffer stream = new StringBuffer();
                long phoneNumber =
ThreadLocalRandom.current().nextlong(99999999501,
                        99999999601);
                int bin = ThreadLocalRandom.current().nextint(1000, 9999);
                int bout = ThreadLocalRandom.current().nextint(1000, 9999);
                stream.append(phoneNumber);
                stream.append(",");
                stream.append(bin);
                stream.append(",");
                stream.append(bout);
                stream.append(",");
                stream.append(new
Date(ThreadLocalRandom.current().nextlong()));

                System.out.println(stream.toString());
                ProducerRecord<Integer, String> data = new
ProducerRecord<Integer, String>(
                        "storm-trident-diy", stream.toString());
                producer.send(data);
                counter++;
            }
            producer.close();
    }
}
```

8.7 小结

在本章，我们阐述了 Trident 的状态以及如何对它进行维护，然后构建了基本的 Trident 拓扑，并带领读者了解了基本的 Trident 拓扑。我们通过示例向读者展示了 Trident 所有可能的操作。此外，还带领读者深入了解了 Trident 内部的工作原理。接着，进一步讲解了 DRPC 在 Trident 中的调用方式及处理过程。最后，通过 `filter`、`group by` 和 `top n` 等操作抛给用户一个问题去自己动手解决。

第四部分
使用 Spark 实现实时计算

- 第 9 章　运用 Spark 引擎
- 第 10 章　运用 Spark 操作
- 第 11 章　Spark Streaming

第 9 章 运用 Spark 引擎

在本章，我们将向读者介绍 Spark 引擎以及 Spark 架构的基础知识，并让读者了解在实际使用中选用 Spark 的必要性和实用性。

本章主要包括以下内容

- Spark 概述
- Spark 的独特优势
- Spark 用例
- Spark 架构——引擎内部的运行模式
- Spark 的语用概念
- Spark 2.x——数据框和数据集的出现

9.1 Spark 概述

Apache Spark 是一个高度分布式计算引擎，其计算的速度与可靠性可以得到保证。Spark 基于 Hadoop 框架，在基于内存计算方面进一步增强，以满足交互式查询和接近实时流处理的需要。并行处理、集群和基于内存的处理等使得 Spark 拥有性能和可靠性方面的优势。Apache Spark 主要的优势如下。

- **速度和效率**：当它运行在传统的基于磁盘的 HDFS 之上时，由于内存计算和磁盘 I/O 的空间节省，速度提高了 100 倍。它将中间结果保存在内存中，从而节省了总执行时间。

- **可扩展性和兼容性**：有多种交互 API 供开发人员选择，包含 Java、Scala 和 Python API 等，以实现开箱即用。
- **分析和机器学习**：为所有机器学习和图计算提供了强大的支持。事实上，在基于大规模数据集开展复杂数据科学研究和人工智能建模等领域，Spark 已经成为开发人员的首选。

Spark 框架和调度器

Spark 框架中的各个组件以及可部署的候选调度器如图 9-1 所示。

图 9-1

图 9-1 包含了 Spark 生态系统的所有基本组件，尽管在一段时间内有些组件已经过时了，但本书在适当的时候也会让用户熟悉这些组件。以下是 Spark 框架的基本组成部分。

- **Spark 内核**：顾名思义，这是 Spark 框架的核心控制单元。主要执行任务的调度和管理。可以使用一种称为**弹性分布式数据集**（Resilient Distributed DataSet，RDD）的 Spark 抽象。RDD 是 Spark 中的基本数据抽象单元，将在下面几节中对它进行详细讨论。

- **Spark SQL**：Spark 的这个模块通常是为能够了解 Oracle、SQL Server、MySQL 等结构化数据集的基础上熟练使用 SQL 的数据工程师而设计的。它支持 Hive，人们可以使用 HiveQL（Hive 查询语言）轻松地查询数据，HiveQL 与 SQL 非常相似。Spark SQL 还可以与 JDBC 和 ODBC 等受用户推崇的连接器进行连接，使开发人员能够将其与前沿的数据库、其他数据集市、BI 以及可视化工具集成。

- **Spark Streaming**：该模块支持实时/近实时流数据处理，并无缝集成到数据管道（如 Kafka、RabbitMQ、Flume 等）中。该模块是为实现可扩展、容错、高速处理而构建的。

- **Spark MLLib**：Spark 的这个模块主要用来实现常用的数据科学统计和机器学习算法。它是一个高度可扩展、分布式和容错的模块，提供了关于分类、成分分析、集群和回归等方面的隐式实现。

- **GraphX**：GraphX 原先是伯克利研究中心的一个独立项目，后来被捐赠给 Spark，成为 Apache Spark 框架的一部分。本质上它是支持图计算和大数据分析的模块，支持 Pregel API 和各种图算法。

- **Spark R**：这是 Spark 最近新增加的模块之一，主要是作为数据科学家的工具而设计的。数据分析师和科学家一直广泛使用 R Studio 作为模型设计工具，但它提供的单节点工具能力有限，因此只能满足数据样本的一个子集。这些模型后面需要在逻辑和优化方面进行大量的返工和重新设计，才能在更广泛的数据集上执行。Spark R 是弥补这一差距的一种尝试，它是一个利用 Spark 功能的轻量级框架，并允许数据科学家在 Spark 引擎中更广泛数据集上，以分布式模式执行 R 模型。

图 9-2 简要概括了 Spark 各种组件的基础功能。

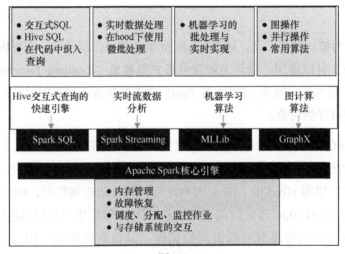

图 9-2

现在我们已经了解了 Spark 框架的基本模块和组件，让我们更深入地了解 Spark 的部署编排机制和设置，有 3 种方式可以部署和协调 Spark，如图 9-3 所示，具体说明如下。

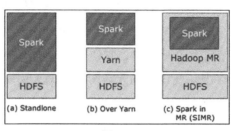

图 9-3

- **Standlone**：在这种 Spark 部署中，Spark 部署在 HDFS 之上。在 HDFS 上为 Spark 分配配额，Spark 作业与其他 MapReduce 作业在相同的 Hadoop Yarn 设置上共同执行。
- **Hadoop Yarn**：这种部署不需要任何预安装，Spark 集成在 Hadoop 生态系统之上——所有作业和 Spark 组件都运行在 Hadoop 生态系统之上。
- **Spark in MapReduce（SIMR）**：我们一直在使用和谈论 Spark on Yarn，但人们一直在努力解决的一个实际挑战是在 V1 版本的 Hadoop 集群上如何运行 Spark 作业。在 Hadoop V1 集群上执行 Spark 作业确实很麻烦，而且需要管理权限。SIMR 是由 Berkley Analytics Lab 联合开发的，因此任何能够访问 Hadoop V1 集群上的 HDFS 和 MapReduce 的用户都可以提交和执行 Spark 作业，无须管理权限或 Spark/Scala 设置。

9.2 Spark 的独特优势

现在大家已经了解了 Spark 组件，接下来进入下一步。在本节，我们将介绍 Spark 在分布式容错处理方面的主要优势，也会谈到 Spark 可能不是最佳解决方案的情况。

- **高性能**：这是 Spark 成功的关键特性，是指基于 HDFS 的高性能数据处理能力。正如在前面看到的，Spark 在 HDFS 和 Yarn 生态系统上使用了它的框架，从而使性能提高了 10 倍，这使得它成为比 map-reduce 更好的选择。Spark 通过限制延迟密集型磁盘 I/O 的使用并利用其内存计算能力来实现这一性能增强。

- **强大且灵活**：Apache Spark 在其开箱即用的实现中是强大的，提供了超过 80 个操作。它是使用 Scala 开发的，并且有 Java、Python 等 API 接口。基础技术和外围技术的全部组合使得它对于任何类型的定制实现都具有很高的可扩展性。

- **内存计算**：内存计算能力是提高 Spark 引擎速度和效率的关键。通过节省写入磁盘的时间来节省总处理时间，它使用一个语用抽象 RDD，并且将大部分的计算存储在内存中。使用**有向无环图**（DAG）引擎进行内存计算和执行流程协调。

- **可重用**：通过一种在批处理、混合处理和实时流处理中可重用的方式并进行一些调整，RDD 抽象帮助程序员在 Spark 中进行开发。

- **容错**：Spark RDD 是 Spark 框架的基本编程抽象，它们具有的弹性不仅体现在这个特性的名称上，实质也是如此。它们被设计成能够在计算过程中处理集群内节点的故障，且不会造成任何数据丢失。虽然 RDD 的设计是为了处理故障，但另一个值得注意的方面是，Spark 利用 HDFS/Yarn 作为其基本框架，Hadoop 在稳定存储方面的弹性是它所固有的。

- **近实时流处理**：Spark Streaming 模块被设计用在流数据之上执行超快的计算和分析，以提供实时的监控和可执行的建议。与 Hadoop 的 MapReduce 相比，这绝对是一个优势，因为在 MapReduce 中，我们不得不等待很长的批处理周期才能得到结果。

- **惰性计算**：Spark 中的执行模型本质上是有惰性的，应用于 RDD 的所有转换都不会立即产生中间结果，而是会形成另一个 RDD。实际执行发生在我们发出操作时。这个模型提高了总执行时间效率。

- **活跃和扩展的社区**：Spark 项目于 2009 年启动，是**伯克利数据分析系统**（BDAS）的一部分。来自 50 多家公司的开发者参与了项目的开发。社区不断扩大，并在业界接受 Spark 的过程中起着至关重要的作用。

- **复杂分析**：是一个设计用于掌握历史批处理记录和实时数据流的处理复杂分析作业的系统。

- **与 Hadoop 集成**：这是 Spark 的另一个性价比高且独特的优势。它与 Hadoop 的集成使其易于被业界采用，因为可以平衡和利用现有的 Hadoop 集群，并能在进行故障保护、缩小范围的情况下提供快速计算。

图 9-4 总结了 Spark 的所有优点。

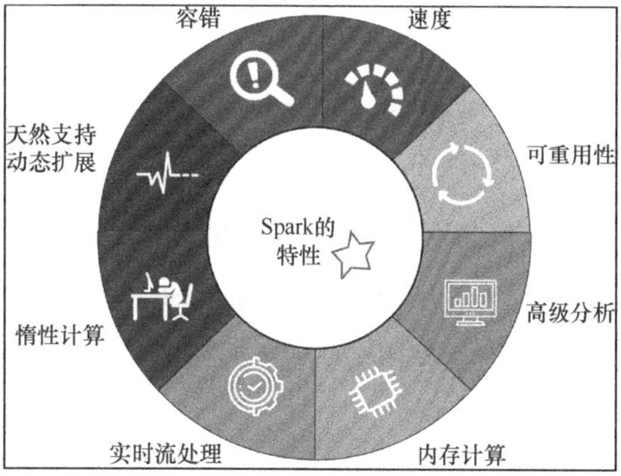

图 9-4

何时避免使用 Spark

前文已经谈及太多关于 Spark 和它的特性，我们不希望用户完全确信它是一个适用于所有类型工具的解决方案，仿佛它是一个神奇的魔杖可以解决用户的所有困难。有些情况下应该避免使用 Spark，下面列出了几种常见的情况。

- **无实时和事件窗口**：Spark 不进行实时处理，实际上是在微批上发挥作用，本质上给人一种 NRT 的错觉。因此，实际是从流事件中创建微批，而不是在一个大的批处理上执行处理逻辑，通过在一个微批上执行，可以得到非常接近实时的结果。这种方法有一个很大的折中，它不能在记录级/事件级上处理断言执行，因为减少批处理大小会对整体性能产生减损影响。

- **小文件和小分区**：如果 Spark 与 Hadoop 在 NRT 中结合使用，可能会遇到这种情况。它在很大程度上取决于任务执行中使用的微批的大小。Hadoop 是专为处理大型/超大型文件块而设计的，而在 Spark 实时组合中，最终会创建大量非常小的文件。对文件读/写处理程序和其他操作系统操作的保留会导致解决方案的性能下降。

- **没有文件系统**：这是一个众所周知的事实，尽管可以认为 Spark 能够灵活地插入任何文件系统（Hadoop、Cloud 等）中。

- **内存比磁盘更昂贵**：Spark 依赖于内存中的计算，因此比 Hadoop-MapReduce 更快，后者使用磁盘操作来实现这一点。但其边际影响是成本的增加，因为内存成本高于磁盘成本。因此，Spark 的性能需要在获取内存和磁盘的额外成本之间进行权衡。

- **延迟**：Spark 提供了比 Hadoop 更低的延迟，但延迟仍然比 Flink 等较新技术要高。
- **背压处理**：背压处理机制不是内置的，它是由开发人员手动处理/构建的。

图 9-5 以图表形式展现了这些情况，以及其他一些类似的情况。其中包括：缺乏优化提示；数据处理的迭代性；开箱即用实现中数据科学算法的数量减少。

图 9-5

现在读者已经清楚地了解了 Spark 的所有显著特性，包括那些不那么耀眼的特性。接下来，我们讨论当 Spark 是分析框架方面最佳选择时的一些用例。

9.3　Spark 用例

本节旨在引导用户了解不同的实际使用用例，其中 Spark 是这些用例解决方案中进行分析处理的最佳和明显选择。

1. 金融领域

- **欺诈识别**：对于所有信用卡用户来说，这是一个非常重要的用例。在这里，通过一系列复杂的数据科学预测算法,可将实时流数据映射到信用卡持卡人和历史使用记录中,从疑似欺诈性信用卡交易中识别出真正的欺诈性交易。据此可为进一步的行动(如允许支付、要求移动验证、阻止交易等)提供依据。

- **Customer 360 模型的客户流失预测及推荐功能**(交叉销售/追加销售):所有金融机构都有大量的数据,但它们在维护数据方面存在困难。如今,我们需要统一地把客户特征和客户行为的正确性实时关联起来,以进一步丰富数据。在使用 Spark 的大型机构中,这种统一的特征化处理非常有效,而且它还有助于在行为数据科学建模中预测客户的流失情况,并基于客户原型和特征化处理方面为交叉销售和追加销售提供建议。

- **实时监控**(更好的客户端服务):实时监控所有渠道的客户活动有助于进行特征化处理和提出相应建议,也有助于监控客户的活动、识别可能出现的异常和违规行为等。例如,同一个客户端不能在 30min 内使用在物理上相距 100mile(1mile=1609.3m)的自动取款机。

2. 电子商务

- **伙伴关系和预测**:许多公司都在使用基于 Spark 的分析来预测市场和进行趋势分析,并相应地建立它们与合作伙伴的基础。

- **阿里巴巴**:它能在 Spark 上运行一些最大规模为数百 PB 的数据分析作业,以支持从图像、商家数据 ETL 和机器学习模型中提取文本,并在这些任务的基础上进行分析、预测、绘制和推荐。

- **制图、分析、ETL、集成**:它为 eBay 做到了这一切。

3. 卫生保健

- **可穿戴设备**:根据由 Spark 使用的实时流数据和患者的既往病史得出的建议来提供临床诊断,由此优化了该行业的配置。它还考虑了其他因素,如国籍、地区饮食习惯、该地区爆发的流行病、天气、温度等。

4. 媒体和娱乐

- **Conviva**:通过删除屏幕缓冲来提供最佳质量的流媒体服务,屏幕缓冲通过实时学习和处理网络问题进行管理，而不会影响服务质量。

- **Netflix**：使用在 Spark 中实现和运行的高级分析算法提供在线推荐。它以 Netflix 系统用户的活动为基础，实现高度的个性化和特征化。

5. 旅行领域

- 个性化推荐（TripAdvisor）。
- **NLP** 和 **Spark** 推荐（OpenTable）。

9.4 Spark 架构——引擎内部的运行模式

前面研究了 Spark 框架的组件、优点/缺点，以及最适合使用它作为解决方案的场景。接下来，我们将深入研究 Spark 的内部结构、架构抽象和工作方式。Spark 在主从模型中工作，图 9-6 显示了该模型的分层体系结构。

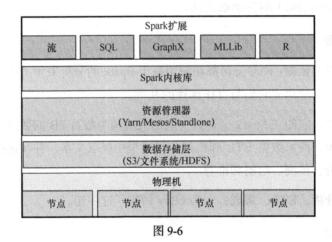

图 9-6

从图 9-6 所示的分层体系结构自下而上开始介绍。

- 物理机或节点由数据存储层（可以是 HDFS/分布式文件系统/AWS S3）抽象。这里的数据存储层提供在存储和检索执行过程中生成的最终/中间数据集的 API。
- 数据存储层之上的资源管理器层使得用户不需要关心与 Spark 安装和执行模型相关的底层存储和资源编排，从而用户可以使用任何可用的资源管理器来配置

Spark，例如 Yarn、Mesos 或 Spark Standlone/Local。

- 在这一层之上，有 Spark 内核和 Spark 扩展层，这在上一节中已经讨论了。

既然已经理解了 Spark 框架在原始物理硬件层上的抽象，那么下一步就是以不同的视角来看 Spark，也就是它的执行模型。图 9-7 展示了 Spark 集群中各个节点下的执行组件。

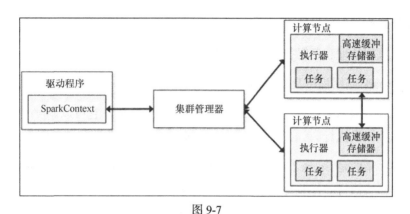

图 9-7

如图 9-7 所示，Spark 集群的主要物理组件包括主节点或驱动程序节点、计算或执行节点和资源管理器。

前面 3 个是执行任何 Spark 应用程序作业必需的关键要素。所有组件的工作方式都与其名称提示相同：**驱动程序节点**承载 Spark 上下文，其中主驱动程序在 Spark 集群中运行；**集群管理器**基本上是 Spark 集群的资源管理组件，而且它可以是 Standlone 资源管理器、基于 Yarn 的资源管理器或基于 Mesos 的资源管理器。它主要处理集群底层资源的协调和管理，以便使用与整体实现无关的方式来执行应用程序。Spark **计算节点**实际上是生成执行程序和任务，并且实际执行 Spark 作业的节点。接下来再仔细看看每个组件。

Spark 集群中的驱动程序是什么？

- 它是在主/驱动程序节点下执行的主进程。

- 它是 Spark Shell 的入口点。

- 它是 Spark 上下文创建的地方。

- RDD 将转换为在驱动程序 Spark 上下文中执行的 DAG。

- 所有任务都是由驱动程序在执行过程中进行调度和控制的。

- RDD 的所有元数据及其沿袭都由驱动程序管理。
- 自带 Spark WebUI。

Spark 集群中的执行器是什么？

- 是创建和执行 Spark 作业任务的从属/计算节点的进程。
- 从外部源读取/写入数据。
- 所有数据处理和逻辑执行都在这里执行。

什么是 Spark 应用程序及其工作组件？

- 是具有数据和计算逻辑的 Spark 上下文的单个实例，它可以调度一系列并行/顺序作业以供执行。
- **作业**：指的是 Spark 作业，是运行在 RDD 上的一组完整的自主转换，通过一个动作完成执行。作业是由驱动应用程序触发的。
- **阶段**：是一组按顺序排列的任务，作为在单个计算节点上执行的管道作业。
- **任务**：是在单个数据分区上执行的一组任务，是 Spark 调度的基本单元。

9.5 Spark 的语用概念

知道开发者最感兴趣的是什么吗？是能够利用框架并根据需求灵活扩展框架。在当今抽象且解耦的世界中，使用各种开箱即用的 API 可以解决这一问题。

在 Spark 出现并将性能提升到新的水平之前，开发者已经讨论了大数据世界一直在努力解决的延迟问题。下面将进一步了解延迟问题。图 9-8 展示了典型的 Hadoop 进程及其中间步骤的执行过程。

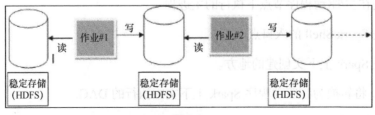

图 9-8

9.5 Spark 的语用概念

如前所述，Hadoop 生态系统广泛利用基于磁盘的分布式稳定存储（HDFS）来存储中间处理结果。

- **作业 #1**：这将从 HDFS 中读取要处理的数据，并将结果写入 HDFS。
- **作业 #2**：这将从 HDFS 中读取作业#1 的临时处理结果，并将结果写入 HDFS。

虽然 HDFS 是一种容错和持久性存储，但从总体延迟来看，任何基于磁盘的读/写操作都是过于奢侈的。要记得 HDFS 的序列化和反序列化，以及 HDFS 的分布式特性都给磁盘读写增加了网络延迟。

因此，虽然这个解决方案是稳健的，但所有这些潜在的延迟增加了作业总周转时间，延迟了最终的处理输出。图 9-9 再次显示了与前面中间步骤中产生存储内存驻留相类似的场景。

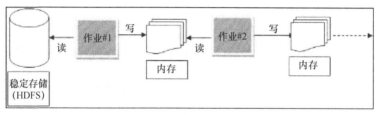

图 9-9

明显的区别在于作业#1 将临时结果写入内存，而作业#2 读取这些中间结果以便在内存中进一步处理，从而节省了磁盘 I/O 操作的时间。

那么，Spark 的神奇之处是什么，使它即使在使用 HDFS 时也能提供更快的结果？神奇之处在于内存中的计算以及**弹性分布式数据集**（Resilient Distributed DataSet，RDD）的抽象。

RDD 是所有 Spark 计算的基本特征和核心，是伯克利大学独立研究的并最初在 Spark 中实施和采用的一个概念。如果必须简单地定义 RDD，可以说它是一种抽象，展示了 Hadoop 的所有特性，但存在于内存中，而不是在磁盘上。

它是一种分布式的、容错的、内存中的数据表示（不可变），可以在分布在 Spark 集群中各个计算节点上的执行器之间，以并行、分布式方式进行进一步的计算。

RDD——词如其名

有弹性的——这意味着 RDD 是容错的。现在的问题是,如果数据在内存中,除非进行了会引起磁盘延迟的持久化,否则怎么把它们从故障中恢复?好吧,在这里告诉所有人——答案是 RDD 逻辑执行计划,它能在任何失败的情况下重新计算丢失/损坏的分区数据。

 RDD 逻辑执行计划实际上是一个 RDD 运算符/依存关系图,能捕获应用于父 RDD 的所有转换的逻辑执行计划,以便在发生故障时可以重新计算整个图/子图以重新生成损坏的 RDD。

分布式——通过集群的不同计算节点,RDD 数据驻留在内存中。

数据集——是一个具有原始值/元组的分区数据集合。

以下是 RDD 的一些更重要功能。

- **内存**:RDD 应尽可能长时间驻留在内存中,以提供更快的访问和低延迟计算。
- **不可变**:当执行了操作并且通过转换生成新的 RDD 时,RDD 是只读的。
- **惰性计算**:支持两种操作,即转换(生成另一个 RDD 的操作)和执行(实际上是触发器和返回值),但所有进程或转换仅在执行操作时以惰性模式来执行。
- **分区**:RDD 中的数据在集群的多个节点之间进行逻辑分区,可以定义位置首选项来计算分区。
- **处理非结构化数据**:这是为了处理非结构化数据而设计的,如文本/媒体流等。

9.6 Spark 2.x——数据框和数据集的出现

我们将介绍 Spark 2.x 中两个新的 Spark 计算抽象。

- **数据框**:分布式的、弹性的、容错的内存数据结构只能处理结构化数据,这意味着它们被设计用来管理以固定类型列划分的数据。虽然对于可以处理任何类型的非结构化数据的 RDD 来说,这听起来像是一个限制,但实际上,这种对数据进行结构化抽象使操作和处理大量结构化数据变得非常容易,就像我们过去使用

RDBMS 方式一样。

- **数据集**：它是 Spark 数据框的扩展，是一种类型安全的面向对象的接口。为简单起见，可以说数据框实际上是一种非类型化的数据集。Spark 语用抽象中最新的 API 实际上使用了 tungsten 内存编码和 catalysts 优化器的特性。

9.7 小结

在本章，我们向读者介绍了 Apache Spark 计算引擎，涉及 Spark 框架的各个组件及调度器，并讨论了 Spark 作为可扩展、高性能的计算引擎拥有的优势。同时读者通过本章也能了解 Spark 不够出色的一些功能和应该避免使用 Spark 的情形。除此之外，本章还向读者介绍了一些实用性强、全行业广泛使用的并紧跟时代发展的用例。然后解析了 Spark 的分层体系结构及其在集群中的内部工作模式。最后，讨论了 Spark 的语用概念：RDD、数据框和数据集。

下一章将讨论 Spark API 以及如何通过计算代码块执行所有组件。

第 10 章 运用 Spark 操作

在本章，我们将向读者介绍 Spark 动作、转换和共享变量，也会介绍 Spark 操作的基本原理，并解释在实际使用中选用 Spark 的必要性和实用性。

本章主要包括以下内容

- Spark——封装和 API
- RDD 语用探索
- 共享变量——广播变量和累加器

10.1 Spark——封装和 API

现在读者已经熟悉了 Spark 的体系结构和基本数据流。在 10.2 节，我们将进一步让读者了解围绕 Spark 构建各种自定义解决方案时经常使用的编程范例和 API。

众所周知，Spark 框架是基于 Scala 开发的，但它也为开发人员提供了使用 Scala、Python 和 Java API 进行交互、开发和定制框架的工具。在下文的讨论中，将仅学习 Scala 和 Java API。

Spark API 可分为两大类：Spark 内核和 Spark 扩展，如图 10-1 所示。

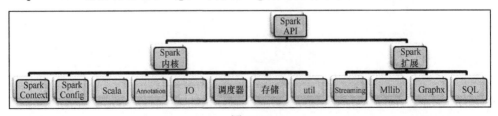

图 10-1

可以看到，在高层次上，Spark 代码库分为两个包。

- **Spark 扩展**：特定扩展的所有 API 都封装在它们自己的包结构中。例如，与 Spark 流相关的所有 API 都在 `org.apache.spark.streaming.*` 包中，其他扩展也有相同的包结构：Spark MLLib 位于 org.apache.spark.mllib.* 中；Spark SQL 位于 org.apcahe.spark.sql.* 中；和 Spark GraphX 位于 org.apache.spark.graphx.* 中。

- **Spark 内核**：Spark 内核是 Spark 的核心，提供了两个基本组件——`SparkContext` 和 `SparkConfig`。每个标准或自定义的 Spark 作业、Spark 库和扩展都使用这两个组件。上下文（context）和配置文件（config）并不是新的术语/概念，它已经或多或少地成为一种标准的体系结构模式。根据定义，上下文是应用程序的入口点，提供对由框架公开的各种资源/功能进行访问，而配置文件包含应用程序配置，这有助于定义应用程序的环境。

作为一名开发人员，读者将会在工作中频繁使用 Spark 内核，因此让我们来看看 Spark 框架的 Spark 内核包中提供的 Scala API 的详细信息。

- `org.apache.spark`：这是 Spark 内核 API 的基本包，它通常封装提供以下功能的类：在集群上 Spark 作业的创建；集群中 Spark 作业的分配与协调；在集群上提交 Spark 作业。

- `org.apache.spark.SparkContext`：这是读者将在任何 Spark 作业/程序中看到的第一个项目。这是 Spark 应用程序的入口点，上下文为开发人员提供了访问和使用 Spark 框架中其他所有功能的途径，以开发和编码业务逻辑驱动的应用程序。它提供了 Spark 作业的执行句柄，甚至提供了对 Spark 扩展的引用。值得注意的是，它是不可变的，每个 JVM 只能实例化一个 Spark 上下文。

- `org.apache.spark.rdd.RDD.scala`：这个封装包含与操作相关的 API，这些操作可以使用分布式 Spark 计算引擎上的 Spark 语用数据单元 RDD 以并行和分布式方式来执行。`SparkContext` 提供了对各种方法的访问，这些方法可从 HDFS/文件系统/Scala 集合中加载数据以创建 RDD。这个包为 RDD 保存各种操作的上下文，比如映射、组合、过滤，甚至是持久化。还有如下一些在某些特定场景中非常有用的专用类。

PairRDDFunctions 在使用键值数据时非常有用；**SequenceFileRDDFunctions** 是处理 Hadoop 序列文件的好助手；**DoubleRDDFunctions** 用于处理有双精度浮点

数的 RDD 函数。

- `org.apache.spark.broadcast`：一旦开始在 Spark 中编程，这将是框架中最常用的包之一，仅次于 RDD 和 `SparkContext`。它封装了在集群中跨 Spark 作业共享变量的 API。本质上，Spark 是用来处理大数据的，本书中所讨论的广播变量的体量是庞大的，因此需要聪明和高效的交换和广播机制，以便在不影响性能和整个作业执行的情况下传递信息。Spark 中有两个广播实现。第一个是 **HttpBroadcast**。顾名思义，此实现依赖于 `HTTPServer` 机制获取和检索服务器本身在 Spark 驱动程序中运行的数据。数据将会存储在执行器和块管理器中。第二个是 **TorrentBroadcast**。这是 Spark 默认的广播实现。在这里，广播机制从执行器/驱动程序中以块的形式获取数据，并保存在自己的块管理器中。原则上，它使用与 BitTorrent 相同的机制来确保驱动程序不会阻碍整个广播管道。

- `org.apache.spark.io`：这里提供了在块存储级别上使用的各种压缩库的实现。整个包被标记为 `DeveloperAPI`，因此可以对其进行扩展，并由开发人员提供自定义实现。默认情况下，它提供 3 种实现：LZ4、LZF 和 Snappy。

- `org.apache.spark.scheduler`：这提供了各种调度器库，以有助于调度、跟踪和监视作业。它定义了**有向无环图**（DAG）调度器。Spark DAG 调度器定义了面向阶段的调度，通过跟踪每个 RDD 的完成情况和每个阶段的输出，计算 DAG，然后进一步提交给在集群上执行它们的 Org.apache.spark.scheduler.taskschedu ler 底层。

- `org.apache.spark.storage`：提供的 API 用于构造、管理并最终将存储在 RDD 块中的数据持久化。它还跟踪数据，并确保将其存储在内存中。如果内存已满，则将数据刷新到底层持久存储区域。

- `org.apache.spark.util`：是跨 Spark API 执行常见功能的实用工具类。它定义了可以用来替代 ScalaTuple2 的 `MutablePair`，但不同的是，`MutablePair` 是可更新的，而 ScalaTuple2 是不可更新的。它有助于优化内存和最大限度地减少对象分配。

10.2 RDD 语用探索

通过前文读者已经理解了 RDD 是一个不可变的、分布式的对象值集合，在 Spark 框架中用作抽象单元。有两种方法可以创建 RDD：加载外部数据集；并行一个已经存在

10.2 RDD 语用探索

于在驱动程序中的列表/集合。

现在来编写一些简单的程序来创建和使用 RDD，如图 10-2 所示。

图 10-2

图 10-2 展示了在 Spark Shell 上创建 RDD 的快速步骤。以下是此操作的具体命令和进一步转换输出。

```
Scala> val inputfile = sc.textFile("input.txt")
```

前面的命令从指定的绝对路径上读取名为 input.txt 的文件，创建一个新的 RDD 并命名为 inputfile。前面的代码片段没有指定完整的路径，因此框架将假定该文件存在于当前目录下。

一旦 RDD 被创建并将前面所提到的输入文本中的数据加载到其中，就能使用它来计算文本中的字数。为此，可以执行以下步骤。

① 以 flatMap 形式，使用空格" "字符将文本划分为多个单词。

② 通过读取每个单词来创建键值对，所有单词的值将都为"1"。

③ 运行 reducer 循环并将相同键的值相加。

Scala 的美妙之处在于，它将前面的所有步骤压缩为一行来执行，如下所示：

```
Scala> val counts = inputfile.flatMap(line => line.split(" ")).map(word =>
(word, 1)).reduceByKey(_+_);
```

下面将分解前面的命令并对其进行简要解释。

- inputfile.flatMap(line => line.split(" "))：在这里，通过空格划分单词来创建 flatMap。

- `map(word => (word, 1))`：使用前面创建的 flatMap，将单词进行映射并赋值为 "1"。
- `(reduceByKey(_+_))`：相加相同键的值。

完成上述的操作后，并不意味着所有操作都已结束。在上一节中已经说过，RDD 在执行方面是有惰性的，现在将真正体验到这一点。之前已经实现了字数转换，但是输出还没有生成，在 Spark 动作之前，将什么也看不到。

现在，应用一个动作并将输出持久化到磁盘。以下命令将新的 RDD 计数持久化到输出文件夹下的磁盘中。

```
Scala> counts.saveAsTextFile("output")
```

读者需要在当前路径下的输出文件夹中查看计数的 RDD 持久化/保存的数据，图 10-3 显示了当前结果。

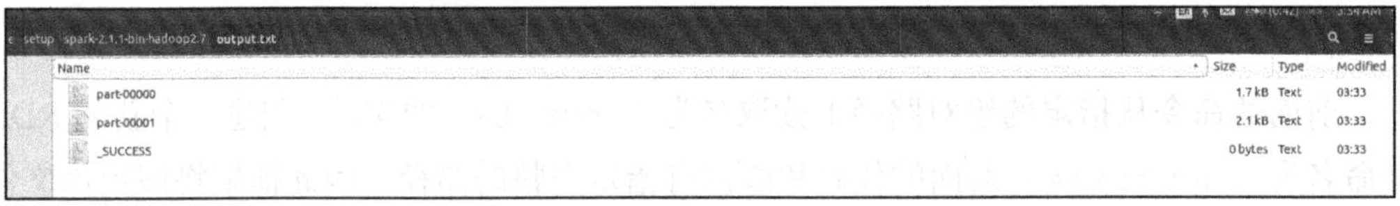

图 10-3

图 10-4 显示了查询的结果。

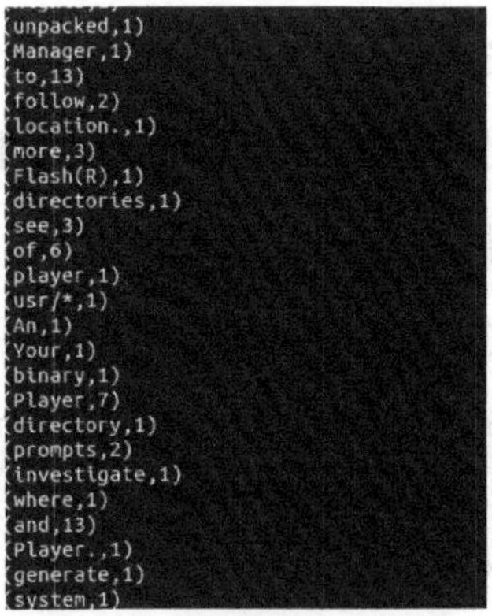

图 10-4

如果还希望从 Web 控制台查看作业的详细信息，可以访问 http://localhost:4040/jobs/，如图 10-5 所示。

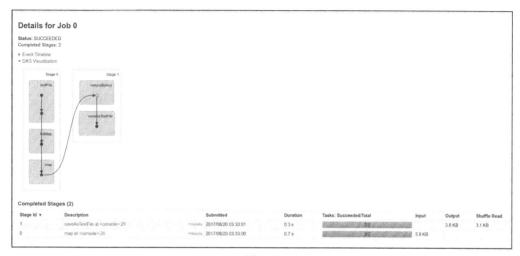

图 10-5

10.2.1 转换

Spark 转换基本上是以一个 RDD 作为输入并输出一个或多个 RDD 的操作。所有转换本质上都是惰性的，而以有向无环图/DAG 形式构建的逻辑执行计划只有在调用一个动作时才会实际执行。

转换可以被划定为窄转换和宽转换。Spark 转换类型见表 10-1。Spark 转换操作的术语解释如图 10-6 和图 10-7 所示。

表 10-1

窄转换	宽转换
窄转换是指来自父 RDD 的每个分区中的数据只用来计算子 RDD 中一个分区的数据，例如 `map()`、`filter()`	宽转换是指使用父 RDD 分区中的数据计算子 RDD 中一个子分区的记录，例如 `GroupBy()`、`ReduceBy()`等
窄转换的操作方式是将计算过程中所需的所有元素存储在父 RDD 的单个分区中	宽转换对存储在父 RDD 分区中的元素或数据记录进行操作

转换	含义
map(*func*)	通过对输入的源数据集中的每一个元素应用函数func，返回一个新的分布式数据集
filter(*func*)	通过函数func选择的元素形成一个新的数据集并返回这个新数据集
flatMap(*func*)	类似于map，但每个输入项都可以映射到0或更多的输出项（因此函数应该返回Seq而不是单个项）
mapPartitions(*func*)	类似于map，但函数单独作用在RDD的每个分区（块）上，因此，当运行类型为T的RDD时，函数的类型必须是Iterator<T> => Iterator<U>
mapPartitionsWithIndex(*func*)	类似于mapPartitions，但应为函数提供表示分区索引的整数值，因此函数当作用于类型为T的RDD时，函数的类型必须为 (Int, Iterator<T>) =>Iterator<U>
sample(*withReplacement, fraction, seed*)	使用给定的随机数生成器的种子，在替换或不替换的情况下，对数据中的一小部分进行采样
union(*otherDataset*)	返回一个新数据集，该数据集包含源数据集与给定数据集中所有元素的集合
intersection(*otherDataset*)	返回一个新的RDD，该RDD包含源数据集和给定数据集中相同元素的集合
distinct([*numTasks*]))	返回一个源数据集去重后的新数据集
groupByKey([*numTasks*])	当被 (K, V) 对形式的数据集调用时，返回 (K, Iterable<V>) 对形式的数据集。注意：1. 如果分组是为了执行聚合（例如对每个键执行求和或求平均值），则使用reduceByKey或aggregateByKey将获得更好的性能。2. 默认情况下，输出中的并行程度取决于父RDD的分区数量。可以传递可选的参数numTask以设置不同数量的任务
reduceByKey(*func*, [*numTasks*])	当被 (K, V) 对形式的数据集调用时，返回一个 (K, V) 对形式的数据集，其中使用给定的归约函数聚合每个键的值，该函数的类型必须是 (V, V) =>V。就像在groupByKey中一样，归约任务的数量可以通过第二个可选参数进行配置

图 10-6

转换	含义
aggregateByKey(*zeroValue*)(*seqOp, combOp,* [*numTasks*])	当被 (K, V) 对形式的数据集调用时，返回 (K, U) 对形式的数据集，其中使用给定的组合函数和一个中性"零"值对每个键的值进行聚合。为避免不必要的分配，允许有不同于输入值类型的聚合值类型。与groupByKey中一样，可以通过第二个可选参数配置减少任务的数量
sortByKey([*ascending*], [*numTasks*])	当被由K实现的有序 (K, V) 对形式的数据集调用时，返回 (K, V) 对形式的数据集，数据集根据键按升序或降序排序，同时ascending参数为布尔类型
join(*otherDataset,* [*numTasks*])	当被类型为 (K, V) 和 (K, W) 的数据集调用时，返回一个 (K, (V, W)) 对形式的数据集，其中包含每个键的所有元素对。外部连接由leftOuterJoin、rightOuterJoin和fullOuterJoin支持
cogroup(*otherDataset,* [*numTasks*])	当被类型为 (K, V) 和 (K, W) 的数据集调用时，返回 (K, (Iterable<V>, Iterable<W>)) 元组形式的数据集。此操作也称为groupWith
cartesian(*otherDataset*)	当被类型为T和U的数据集调用时，返回 (T, U) 对（所有元素对）形式的数据集
pipe(*command,* [*envVars*])	通过shell命令（如Perl或bash脚本）对RDD的每个分区调用pipe。RDD元素被写入进程的stdin，结果以字符串形式的RDD返回到stdout
coalesce(*numPartitions*)	将RDD中的分区数减少到NumPartitions。这对于更高效地过滤大型数据集十分有用
repartition(*numPartitions*)	随机重排RDD中的数据，以创建更多或更少的分区，并在这些分区之间保持平衡。这通常会将网络上的所有数据都进行重排
repartitionAndSortWithinPartitions(*partitioner*)	根据给定的分区器对RDD进行重新分区，并在每个结果分区中，按其键对记录进行排序。这比调用repartition然后在每个分区内进行排序更有效，因为它可以在shuffle过程中进行排序

图 10-7

10.2 RDD 语用探索

转换的分类如图 10-8 所示。

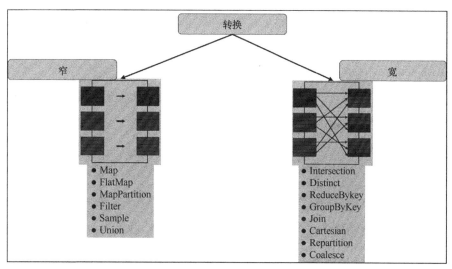

图 10-8

以下部分展示了各种转换操作的代码段。

- Map：以下代码读取名为 map_spark_test.txt 的文件，将文本的每一行映射到其长度，并打印出来。

```
import org.apache.spark.SparkContext
import org.apache.spark.SparkConf
import org.apache.spark.sql.SparkSession
object mapOperationTest{
def main(args: Array[String]) = {
    val sparkSession =
SparkSession.builder.appName("myMapExample").master("local").ge
tOrCreate()
    val fileTobeMapped = sparkSession.read.textFile("my
_spark_test.txt").rdd
    val fileMap = fileTobeMapped.map(line =>
(line,line.length))
    fileMap.foreach(println)
    }
}
```

- flatMap()：Map 和 flatMap 操作类似，两者都应用于输入 RDD 中的每个元素，但方式不同。因为 Map 返回单个元素，flatMap 可以返回多个元素。

在下面的代码段中，将执行操作 flatMap 来划分输入文本中每一行的单词。

```
val fileTobeMapped =
spark.read.textFile("my_spark_test.txt").rdd
val flatmapFile = fileTobeMapped.flatMap(lines => lines.split("
"))
flatmapFile.foreach(println)
```

- **Filter** 操作：只返回来自父 RDD 且满足过滤器所提条件的元素。以下代码段计算其中包含单词 spark 的行数。

```
val fileTobeMapped = spark.read.textFile("my_spark_test.txt").rdd
val flatmapFile = fileTobeMapped.flatMap(lines => lines.split("
")).filter(value => value=="spark")
println(flatmapFile.count())
```

- union：此转换接受两个或多个相同类型的 RDD，并输出包含这两个 RDD 元素的 RDD。

以下代码片段合并了 3 个 RDD 并打印了输出的 RDD。

```
val rddA =
spark.sparkContext.parallelize(Seq((2,"JAN",2017),(7,"NOV",2015
),(16,"FEB",2015)))
val rddB =
spark.sparkContext.parallelize(Seq((6,"DEC",2015),(18,"SEP",201
6)))
val rddC =
spark.sparkContext.parallelize(Seq((7,"DEC",2012),(17,"MAY",201
6)))
val rddD = rddA.union(rddB).union(rddC)
rddD.foreach(Println)
```

- intersection：在相同类型的两个或多个 RDD 之间运行的一种操作，用于返回在所有输入的 RDD 中共有的元素。

```
val rddA =
spark.sparkContext.parallelize(Seq((2,"JAN",2017),(4,"NOV",2015
),(17,"FEB",2015)))
val rddB=
spark.sparkContext.parallelize(Seq((5,"DEC",2015),(2,"JAN",2017
)))
val comman = rddA.intersection(rddB)
comman.foreach(Println)
```

- `distinct`：此转换有助于从父 RDD 中删除重复项。

```
val rddA = 
park.sparkContext.parallelize(Seq((2,"JAN",2017),(4,"NOV",2015)
,(17,"FEB",2015),(3,"NOV",2015)))
val result = rddA.distinct()
println(result.collect().mkString(", "))
```

- `groupByKey()`：当对键值数据集应用此转换时，将导致数据根据键进行重排。这是一个高度网络密集型的操作，为提高效率，应注意缓存和磁盘持久性。

下面的代码片段根据字母对数字进行分组，其中字母用作键，数字是值。函数 collect() 将其作为数组返回。

```
val mydata = 
spark.sparkContext.parallelize(Array(('k',5),('s',3),('s',4),('
p',7),('p',5),('t',8),('k',6)),3)
val group = mydata.groupByKey().collect()
group.foreach(println)
```

- `reduceByKey()`：此操作适用于基于键值的数据集，在重排数据集中的数据之前，在同一台计算机上而不是通过网络，相同键的值将被映射并被分组在一起。

```
val myWordList = 
Array("one","two","two","four","five","six","six","eight","nine
","ten")
val myWordList = spark.sparkContext.parallelize(words).map(w =>
(w,1)).reduceByKey(_+_)
myWordList.foreach(println)
```

- `sortByKey()`：该操作适用于基于键值对的数据，其中这些值按键的顺序进行排列。

```
val myMarkList = 
spark.sparkContext.parallelize(Seq(("maths",52),
("english",75), ("science",82), ("computer",65), ("maths",85)))
val mySortedMarkList = myMarkList.sortByKey()
mySortedMarkList.foreach(println)
```

- `join()`：这是一种类似于数据库中连接的转换。在 Spark 中，它的操作对象是成对的 RDD，RDD 中每个元素都是一个元组——第一个元素是键，第二个元素是值。组合操作接收两个输入的 RDD 并根据键连接它们，生成输出 RDD。

```
val rddA = 
```

```
spark.sparkContext.parallelize(Array(('A',1),('B',2),('C',3)))
val rddB
=spark.sparkContext.parallelize(Array(('A',4),('A',6),('B',7),(C',3),('C',8)))
val result = rddA.join(rddB)
println(result.collect().mkString(","))
```

- `coalesce()`:这是一个非常有用的操作,用于通过控制为数据集定义的分区数量来减少跨 Spark 集群中节点的数据重排。基于合并操作定义分区数量,数据将分布在多个节点上。

在以下代码段中,来自 `rddA` 的数据将仅分布在 2 个分区的 2 个节点上,即使 Spark 集群共有 6 个节点也是如此。

```
val rddA =
spark.sparkContext.parallelize(Array("jan","feb","mar","april","may","jun"),3)
val myResult = rddA.coalesce(2)
myResult.foreach(println)
```

10.2.2 动作

简而言之,可以说动作实际上是对真实数据执行转换以生成输出。动作将数据从执行器发送到驱动程序。图 10-9 展示了动作的返回值。

下面来看一些代码片段,看看动作的实际执行。

- `count()`:顾名思义,这个操作计算并提供了 RDD 中的元素数量。在前面的过滤器示例中,使用了计数操作,该操作计算正在读取的文本中每行的字数。
- `collect()`:收集操作如名称所示,将 RDD 中的所有数据返回给驱动程序。由于此操作的性质,建议读者谨慎使用它,因为它将数据复制到驱动程序,所以所有数据都应该适合驱动程序正在执行的节点。

在下面的代码片段中,根据字母的键将这两个 RDD 连接在一起,然后把生成的 RDD 返回到驱动程序。在驱动程序中,连接的值作为与键对应的数组进行保存。

```
val rddA =
spark.sparkContext.parallelize(Array(('A',1),('b',2),('c',3)))
val rddB
=spark.sparkContext.parallelize(Array(('A',4),('A',6),('b',7),('c',3),('c',8)))
val resultRDD = rddA.join(rddB)
```

10.2 RDD 语用探索

```
println(result.collect().mkString(","))
```

动作	含义
reduce(*func*)	使用函数func（接收两个参数并返回一个值）聚合数据集中的元素。该函数应该是满足交换律和结合律的，以便能够正确地执行并行计算
collect()	在驱动程序中以数组的形式返回数据集中的所有元素，这通常在执行过滤或其他操作得到足够小的数据子集时非常有用
count()	返回数据集中的元素数
first()	返回数据集的第一个元素（类似于`take (1)`）
take(*n*)	返回有数据集前n个元素的数组
takeSample(*withReplacement, num, [seed]*)	以替换或不替换，或者预先指定随机数生成器种子的形式，返回一个数组，该数组包含数据集中num个元素的随机样本
takeOrdered(*n, [ordering]*)	根据RDD的原有顺序或自定义比较器返回RDD中前n个元素
saveAsTextFile(*path*)	将数据集中的元素作为文本文件（或一组文本文件）写入本地文件系统、HDFS或任何其他Hadoop支持的文件系统的给定目录中。Spark将调用每个元素的String，将其转换为文件中的一行文本
saveAsSequenceFile(*path*) (Java and Scala)	将数据集中的元素作为Hadoop SequenceFile写入本地文件系统、HDFS或任何其他Hadoop支持的文件系统的给定路径中。这在实现了Hadoop的Writable接口的键值对在RDD上可用。在Scala中，它也适用于隐式转换为Writable的类型（Spark包含了Int、Double、String等基本类型的转换）
saveAsObjectFile(*path*) (Java and Scala)	使用Java序列化以一种简单格式编写数据集中的元素，并且可以使用`SparkContext.objectFile()`加载
countByKey()	仅在 (K, V) 类型的RDD上可用。返回 (K, Int) 对的键对应计数的一个散列映射
foreach(*func*)	对数据集中的每个元素应用函数func。这通常用于副作用，如更新Accumulator或与外部存储系统交互。 注意：修改`foreach()`之外的Accumulator以外的变量可能会导致未定义的行为

图 10-9

- `top()`：此操作按照默认或指定的排序规则返回 RDD 前面的元素。这有利于采样。以下代码片段根据文件的默认顺序返回文件中的前 3 条记录。

```
val fileTobeMapped =
spark.read.textFile("my_spark_test.txt").rdd
val mapFile = fileTobeMapped.map(line => (line,line.length))
val result = mapFile.top(3)
result.foreach(println)
```

- `countByValue()`：此操作返回元素在输入 RDD 中出现的次数。其输出采用键值对的形式，其中键是元素，值表示其计数。

```
val fileTobeMapped = spark.read.textFile("my_spark_test.txt
").rdd
val result= fileTobeMapped.map(line =>
```

```
(line,line.length)).countByValue()
result.foreach(println)
```

- `reduce()`：此操作从输入 RDD 中提取两个元素，输出的类型与输入 RDD 保持相同——操作结果可以是计数值，也可以是相加值，具体取决于传递的参数。在执行方面，这个动作的执行是满足结合律与交换律的。

以下代码片段对输入 RDD 中的元素进行了求和。

```
val rddA =
spark.sparkContext.parallelize(List(20,32,45,62,8,5))
val sum = rddA.reduce(_+_)
println(sum)
```

- `foreach()`：这个动作的工作方式正如其名称所示——应用于所操作的 RDD 的每个元素。

在下面的代码片段中，根据键（字母）对 `rddA` 进行分组，使用 `collect` 操作执行转换，并将开放的 `println` 函数应用于 `myGroupRdd` 的每个元素中，从而打印在控制台上。

```
val rddA =
spark.sparkContext.parallelize(Array(('k',5),('s',3),('s',4),('p',7),('p',5),('t',8),('k',6)),3)
val myGroupRdd = data.groupByKey().collect()
myGroupRdd.foreach(println)
```

10.3　共享变量——广播变量和累加器

当在分布式计算程序和模块中工作时，代码在不同的节点和/或不同的计算节点上执行，很多时候需要在分布式执行设置的执行单元之间共享数据。因此 Spark 有共享变量这个概念。共享变量用于在不同计算节点中的并行执行任务之间，或任务与驱动程序之间共享信息。Spark 支持两种类型的共享变量：广播变量和累加器。

接下来，我们将从概念和实用性两个方面研究这两种类型的 Spark 变量。

10.3.1　广播变量

广播变量是程序员打算在整个集群中与所有执行单元共享的变量。虽然这听起来很简单,但在使用广播变量时需要了解以下几个方面:广播变量需要能够适配集群中每个节点的内存——就像是每个节点的本地只读字典/索引,它们的大小不能太大,并且所有节点共享相同的值,所以在设计上它是只读的。例如,我们有一个用于拼写检查的字典,那么希望每个节点都有相同的副本。

综上所述,以下是设计和使用广播变量的主要注意事项/特性:广播变量是不变的;广播变量分布在整个集群中(计算节点和驱动程序);广播变量必须与内存相适应。

广播变量非常适合静态查找表或元数据。在这些表或元数据中,每个节点都有自己的共享副本,并且不需要将数据发送到每个节点,从而节省了大量的网络 I/O。

图 10-10 展示了广播变量的显式声明和用法,其中包含值 1、2 和 3。

```
scala> val broadcastVar = sc.broadcast(Array(1, 2, 3))
broadcastVar: org.apache.spark.broadcast.Broadcast[Array[Int]] =
    Broadcast(0)

scala> broadcastVar.value
res0: Array[Int] = Array(1, 2, 3)
```

图 10-10

下面将展示一个小的代码示例,以演示 Scala 代码中相同的用法。以下是包含印度一些州名称的 CSV:

```
Uttar Pradesh, Delhi, Haryana, Punjab, Rajasthan,
Maharashtra, Tamilnadu, Telangana
```

接下来,从磁盘将其加载到映射中,然后将其转换为广播变量。

```
def loadCSVFile( filename: String): Option[Map[String, String]]
    val states= Map[String, String]()
    Try {
    val bufferedSource = Source.fromFile(filename)
    for(line <- bufferedSource.getLines) {
    val Array(state, capital) = line.split(","),map(_.trim)
    states +=state -> capital
    }
    bufferedSource.close()
    return Some(states)
    }.toOption
}
```

第10章 运用 Spark 操作

在下一步中,把这个映射转换成一个广播变量。

前面的代码片段已加载了州名称的文件,并将其转换为作为 statesCache 广播的映射。接下来,将由映射州的键创建一个 stateRDD,使用 searchStateDetails 方法从用户指定的特定字母表中搜索州名,并返回它的详细信息(如首府等)。

在这种机制中,不需要在每次执行搜索操作时都将州的 CSV 文件发送给每个节点和执行器。

从下面的代码片段中可以看到,前面引用示例的完整源代码。

```
import org.apache.spark.{ SparkContext, SparkConf }
import org.apache.spark.rdd.RDD
import org.apache.spark.broadcast.Broadcast

import scala.io.Source
import scala.util.{ Try, Success, Failure }
import scala.collection.mutable.Map

object TestBroadcastVariables {
    def main(args: Array[String]): Unit = {

        loadCSVFile("/myData/states.csv") match {
          case Some(states) => {
                val sc = new SparkContext(new SparkConf()
                .setAppName("MyBroadcastVariablesJob"))

                val statesCache = sc.broadcast(states)
            val statesRDD = sc.parallelize(states.keys.toList)

            // happy case...
            val happyCaseRDD = searchStateDetails(statesRDD,
statesCache, "P")
                println(">>>> Search results of states starting with 'P': "
+ happyCaseRDD.count())
                happyCaseRDD.foreach(entry => println("State:" + entry._1 +
", Capital:" + entry._2))

                // non-happy case...
                val nonHappyCaseRDD = searchStateDetails(statesRDD,
statesCache, "Yz")
                println(">>>> Search results of states starting with 'Yz': 
" + nonHappyCaseRDD.count())
                nonHappyCaseRDD.foreach(entry => println("State:" +
```

```
            entry._1 + ", Capital:" + entry._2))
            }
            case None => println("Error loading file...")
        }
    }

def searchStateDetails(statesRDD: RDD[String], stateCache:
Broadcast[Map[String, String]],
    searchToken: String): RDD[(String, String)] = {
        statesRDD.filter(_.startsWith(searchToken))
        .map(country => (state, stateCache.value(state)))
}
def loadCSVFile(filename: String): Option[Map[String, String]] = {
    val countries = Map[String, String]()
    Try {
        val bufferedSource = Source.fromFile(filename)

        for (line <- bufferedSource.getLines) {
        val Array(state, capital) = line.split(",").map(_.trim)
        states += state -> capital
        }
    bufferedSource.close()
    return Some(states)

        }.toOption
    }
}
```

10.3.2 累加器

这是在 Spark 作业中跨不同节点或驱动程序共享值/数据的第二种方法。从这个变量的名称中可以明显看出，累加器用于计数或累加值。累加器源于 MapReduce 的 counters，不同于广播变量，它们是可变的——累加器的值可以更改，作业可以更改累加器的值，但只有驱动程序可以读取它的值。累加器在 Spark 的分布式计算节点中充当数据聚合和计数的好帮手。

假设有一个商店 Walmart 的购买日志，我们需要编写一个 Spark 作业来检测日志中每种不良记录类型的数量。下面的代码片段将实现这一功能。

```
def main(args: Array[String]): Unit = {

val tnxt = new SparkContext(new SparkConf().setAppName("SaleAnalysisJob"))
```

```
val badtnxts = tnxt.accumulator(0, "Bad Transaction")
val zeroValuetnxt= tnxt.accumulator(0, "Zero Value Transaction")
val missingFieldstnxt = tnxt.accumulator(0, "Missing Fields Transaction")
val blankLinesTnxt = tnxt.accumulator(0, "Blank Lines Transaction")

ctx.textFile("file:/mydata/sales.log", 4)
    .foreach { line =>
        if (line.length() == 0) blankLines += 1
        else if (line.contains("Bad Transaction")) badtnxts+= 1
        else {
            val fields = line.split("\t")
            if (fields.length != 4) missingFieldstnxt+= 1
            else if (fields(3).toFloat == 0) zeroValuetnxt += 1
        }
}
println("Sales Log Analysis Counters:")
println(s"\tBad Transactions=${ badtnxts.value}")
println(s"\tZero Value Sales=${ zeroValuetnxt.value}")
println(s"\tMissing Fields Transactions=${ missingFieldstnxt.value}")
println(s"\tBlank Lines Transactions=${ blankLines.value}")
```

10.4 小结

在本章，我们向读者介绍了 Apache Spark API 及其组织架构，在理论上和实例中讨论了转换和动作的概念，并带领读者进入了共享变量的领域：广播变量和累加器。在第 11 章，我们专门讨论 Spark Streaming。

第 11 章 Spark Streaming

本章将介绍 Spark Streaming、体系结构和微批的概念；将研究流应用的各种组件，以及集成了大量输入源的流应用的内部结构；还将进行一些实践练习，以演示动作中流应用的执行。

---- 本章主要包括以下内容 --

- Spark Streaming 的概念
- Spark Streaming 的简介和体系结构
- Spark Streaming 的封装结构
- 连接 Kafka 和 Spark Streaming

11.1 Spark Streaming 的概念

Spark 框架及其所有扩展一起提供了一种通用的解决方案，该方案可以满足批处理、分析和实时等企业数据需求。为了能够执行实时数据处理，该框架应该能够处理近乎实时的无界数据流。此功能是由 Spark 框架中 Spark Streaming 扩展下的微批和流处理提供。

简单来说，可以将数据集理解为一个不断实时生成的无界数据序列。现在，为了能够处理这些不断到达的数据流，各种框架对它们的处理方式如下。

- 单独处理的不同的离散事件。
- 将单个事件通过微批变为非常小的批次，这些批次作为单个单元进行处理。

Spark 提供流 API 作为其内核 API 的扩展，内核 API 是一个可扩展、低延迟、高吞吐量和容错的框架，能使用微批实时处理传入的流数据。

在某些用例中，基于 Spark 框架的实时处理方案将派上用场。这些用例包括：监控基础设施、应用程序或进程；检测欺诈；营销和广告；物联网。

图 11-1 截取了世界各地关于实时数据生成速率的一些统计数据，其中描述的所有场景往往都可作为 Spark Streaming 处理的用例。

图 11-1

11.2 Spark Streaming 的简介和体系结构

Spark Streaming 是 Spark 内核 API 的一个非常有用的扩展，被广泛用于处理实时或接**近实时**传入的流数据，就像在**近实时**（**NRT**）中一样。此 API 扩展具有所有 Spark 内核功能，即高度分布式、可扩展、容错、高吞吐量和低延迟处理。

图 11-2 显示了 Spark Streaming 如何与 Spark 执行引擎紧密地配合工作，以处理实时数据流。

图 11-2

Spark Streaming 工作在基于微批的体系结构上——可以将其设想为内核 Spark 体系

结构的扩展。在内核 Spark 体系结构中，该框架通过将流中的传入事件分类到确定的批中来执行实时处理。每个批次的大小相同，实时数据被收集并堆叠到这些确定大小的微批中进行处理。

在 Spark 框架下，使用用户定义的批处理持续时间确定每个微批的大小。为了更好地理解它，让我们举一个例子。假设程序每秒接收 20 个事件的实时/流式数据，用户提供的批处理持续时间为 2s。现在，Spark Streaming 将不断处理到达的数据，但它将在每 2s 结束时对收到的数据创建微批（每个批处理包含 40 个事件），并将其提交给用户定义的作业进行进一步处理。

开发人员需要了解的一个重要方面是，从事件发生到结果生成的整个执行周期内批处理大小/持续时间和进程延迟成反比。批处理大小通常基于以下两个标准来定义：性能与优化；业务/用例可接受的延迟。

这些微批被称为 Spark Streaming 中的**离散化流**（Discretized Stream，DStream），是**一系列弹性分布式数据集**（RDD）。

在本节，我们将让读者了解 Spark Streaming 体系结构不同于传统流处理体系结构的主要方面，以及这种基于微批的处理抽象相对于传统的基于事件的处理系统的优点。这种对比会让读者了解 Spark Streaming 体系结构和设计的内在本质。

传统的流系统一般都是基于事件的，当数据到达时，系统中的各种算子都会对其进行处理。分布式设置中的每个执行单元一次处理一条记录，如图 11-3 所示。

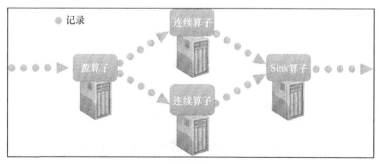

图 11-3

图 11-3 捕捉了典型的模型，其中，从源中一次提取一条记录并传播到计算节点，在该节点上逐条处理记录，Sink 算子将其转储到稳定存储或下游节点以进一步处理。

第 11 章 Spark Streaming

在继续下一步之前,需要明确基于 Storm 的流处理引擎和基于 Spark 的流处理平台的主要区别。图 11-4 清楚地展示并阐明了这两个顶级流处理平台的各种功能。

这两种框架的根本区别在于 Storm 对任务并行计算,而 Spark 平台对数据并行计算,如下所示。

- **任务并行**:这种形式的并行性包括在同一台或多台计算机上跨多个处理器/计算节点线程执行任务/程序单元。它侧重于并行执行不同的操作,以充分利用处理器和内存的可用计算资源。
- **数据并行**:这种并行形式主要是数据集在多个计算程序之间的分布。这种形式中,在分布式数据子集的不同并行计算处理器中执行相同的操作。

特性	Apache Storm/Trident	Spark 流
编程语言	Java Clojure Scala	Java Scala
可靠性	支持"精确一次"处理模式。也可用于其他模式,如"至少一次""至多一次"处理模式	支持"精确"处理模式
流处理源	Spout	HDFS
基元流	Tuple, Partition	DStream
持久化	MapState	Per RDD
状态管理	Supported	Supported
资源管理	Yarn, Mesos	Yarn, Mesos
配置	Apache Ambari	Basic monitoring using ganglia [3]
消息传递	ZeroMQ, Netty	Netty, Akka

图 11-4

Storm Trident 和 Spark 都提供基于时间窗口限制的流和微批功能。虽然在功能特性上它们相似,但在实现语义上有所不同,选择哪种框架,取决于应用/问题的需求。图 11-5 阐明了一些基本规则。

在前面几点的基础上,下面是一些最终的注意事项。

- **延迟**:Storm 可以轻松地提供亚秒级延迟,但对于 Spark Streaming 来说,实现这一点并不容易,因为它本质上不是一个流平台,而是一个微批。
- **总拥有成本**(TCO):对于任何应用来说,这都是一个非常重要的考虑因素。如果应用程序在批处理和实时方面需要类似的解决方案,则 Spark 具有优势,因为它可以使用相同的应用代码库,从而节省开发成本。但 Storm 中的情况并非如此,

11.2 Spark Streaming 的简介和体系结构

因为它与 **MapReduce**/Batch 应用的实现有很大不同。

情景	框架的选择
严格低延迟	与Spark流相比，Storm可以以更少的限制提供更低的延迟
低开发成本	在Spark中，相同的代码库可以用于批处理和流处理。但在Storm中，这是不可能的
消息传递保证	Apache Storm（Trident）和Spark流都支持"精确一次"处理模式
容错	这两个框架在相同程度上都是相对容错的。 在Apache Storm/Trient中，如果进程失败，当状态管理通过Zookeeper进行处理时，监管者进程将自动重新启动该进程。 Spark通过资源管理器（可以是Yarn、Mesos或Standlone）重新启动计算节点

图 11-5

- **消息传递**：Spark 中默认是"精确一次"语义，而 Storm 提供至少一次且恰好一次。在 Storm 中实现后者实际上有点棘手。

图 11-6 描述了各种体系结构的组件，如**输入数据流**、**输出数据流/存储**等。在 Spark Streaming 程序的执行过程中，这些组件具有关键作用和自己的生命周期。

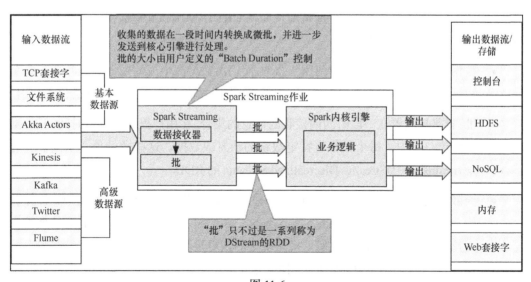

图 11-6

接下来看一下图 11-6 中描述的组件。

（1）**输入数据流**：这里定义的输入数据源，本质上是以非常高的频率（秒级、毫秒）传送实时/流式数据的源。这些源可以是原始套接字、文件系统，甚至可以是高度可扩展的队列产品，例如 Kafka。Spark Streaming 作业使用各种可用连接器连接到输入数据源。这些连接器可能由 Spark 发行版本身提供，也可能需要单独下载并在 Spark Streaming 作业中进行配置。这些输入流也称为 **Input DStream**。根据连接器的可用性，输入数据源分为以下类别。

- **基本数据源**：基本数据源的连接器及其所有依赖项都封装在 Spark 的标准发行版中，不需要另外下载任何其他软件包。

- **高级数据源**：为了避免复杂性和版本冲突，高级数据源的连接器及其所需的依赖项在 Spark 的标准发行版中不可用。因此需要分别下载并配置这些连接器的依赖项，或者按照每个数据源的集成指南（见 Kafka、Flume 和 Kinesis 的相关网站）中的说明，将它们作为 Maven 脚本的依赖项。

- 有关可用的高级数据源的列表，请参阅相关网站。

> Spark 社区还提供了各种其他连接器，这些连接器可以直接从 Spark 官网上下载，也可以参考相关网站来开发定制连接器。

（2）**Spark Streaming 作业**：这是用户为在 NRT 中使用和处理数据反馈而开发的自定义作业。由以下几个方面组成。

- **数据接收器**：这是一个专门用于接收/使用由数据源生成的数据的接收器。每个数据源都有自己的接收器，而且不能在不同类型的数据源之间进行泛化或共用。

- **批**：批是接收器在一段时间内接收的消息的集合。在用户提供的特定时间间隔（批处理窗口）内，每个批都有特定数量的消息或数据。这些微批是一系列被称为 **Dstreams**——DStreams-Discrealized Streams 的 RDD。

- **离散化流**：也称为 DStream，这是一种新的流处理模型，在这种模型中，计算被构造为一系列时间间隔小的、无状态的、确定性的批计算。这种新的流处理模式支持强大的恢复机制（类似于批系统中的恢复机制），并且性能优于复制和上游备份。离散化流扩展和利用了弹性分布式数据集的概念，并在单独的 DStream 中创建了一系列（相同类型的）RDD，该 DStream 是按照用户定义的时间间隔（批处理持续时间）进行处理和计算的。DStream 可以从连接到不同数据源（如套接字、文件系统等）的输入数据流中进行创建，也可以通过在其他 DStream 上应用高级操作来创建（类似于 RDD）。每个 Spark Streaming 的上下文可以有多个 DStream，并且每个 DStream 都包含一系列 RDD。每个 RDD 都是在特定时间点从接收器上接收的数据的快照。

- **流上下文**：Spark Streaming 扩展了 Spark 上下文，并提供了一个新的上下文，流上下文用于访问 Spark Streaming 的所有功能和特性。流上下文是主要入口点，并提供了从各种输入数据源上初始化 DStream 的方法。

- **Spark 内核引擎**：内核引擎以 RDD 的形式接收输入，并根据用户定义的业务逻辑进一步处理，最后将其发送到关联的**输出数据流/存储**。

- **输出数据流/存储**：每个被处理的批处理的最终输出都会被发送到输出流以进行进一步的处理。这些输出数据流可以是不同类型的，包括原始文件系统、WebSocket、NoSQL 等。

11.3 Spark Streaming 的封装结构

在本节，我们将讨论 Spark Streaming 中公开的各种 API 和操作。

11.3.1 Spark Streaming API

所有 Spark Streaming 类都封装在 `org.apache.spark.streaming.*` 包中。Spark Streaming 定义了两个核心类，这两个类提供对所有 Spark Streaming 功能的访问，例如 `StreamingContext.scala` 和 `DStream.scala`。

让我们研究一下这些类的功能和扮演的角色。

- `org.apache.spark.streaming.StreamingContext`：这是 Spark Streaming 功能的入口点。它定义了创建 `DStream.scala` 对象以及启动和停止 Spark Streaming 作业的方法。

- `org.apache.spark.streaming.dstream.DStream.scala`：DStream 或离散化流为 Spark Streaming 提供了基本抽象。它们提供了从实时数据上创建的

RDD 序列，用于转换现有的 DStream。此类定义了可在所有 DStream 上执行的全局操作，以及可应用于特定 DStream 类型的一些特定操作。

除了前面定义的类，Spark Streaming 还定义了各种子包，为各种类型的输入接收器提供功能，如下所示。

- `org.apache.spark.streaming.kinesis.*`：这个包提供了一些类，用于处理来自 Kinesis 的输入数据。
- `org.apache.spark.streaming.flume.*`：这个包提供了一些类，用于处理来自 Flume 的输入数据。
- `org.apache.spark.streaming.kafka.*`：这个包提供了一些类，用于处理来自 Kafka 的输入数据。
- `org.apache.spark.streaming.zeromq.*`：这个包提供了一些类，用于处理来自 ZeroMQ 的输入数据。
- `org.apache.spark.streaming.twitter.*`：这个包提供了一些类，用于处理来自使用 twitter4j 的 Twitter 订阅源的输入数据。

11.3.2　Spark Streaming 操作

Spark 提供了可以在 DStream 上执行的各种操作，所有操作都分为转换和输出操作。让我们来讨论一下这两个操作。

（1）**转换操作**：转换是用于修改或更改输入流中数据结构的操作。所有转换都是相似的，并且支持 RDD 提供的几乎所有转换操作，例如 `map()`、`platmap()`、`union()` 等。除了 DStream 定义的常规转换操作，与 RDD 类似，DStream 还对流数据提供了一些特殊的转换操作。让我们来讨论一下这些操作。

- **窗口操作**：窗口是一种特殊类型的操作，仅由 DStream 提供。窗口操作将过去时间间隔内滑动窗口中的所有记录分组到一个 RDD 中。窗口操作提供了对需要进行分析和处理的数据定义范围的功能。DStream API 还在滑动窗口上提供了增量聚合或处理功能，可以在滑动窗口上计算聚合，例如，计数或最大值。DStream API 提供了各种窗口操作。**DStream.scala** 中的所有以 **Window** 为前缀的方法都提供增量聚合，如 **countByWindow**、**reduceByWindow** 等。

- **转换操作**：转换操作（如 **transform(....)** 或 **transformWith(...)**），是一种特殊类型的操作，提供了执行任意 RDD 到 RDD 操作的灵活性。从本质上说，这有助于执行 DStream API 没有提供/公开的任何 RDD 操作。此方法还用于合并两个 Spark 领域，即批和流。也就是说，可以使用批处理创建 RDD，并与使用 Spark Streaming 创建的 RDD 进行合并。它有利于跨 Spark 批和流的代码重用，比如在 Spark 批处理应用程序中可以编写函数，并且希望在 Spark Streaming 应用中使用这些函数。

- **updateStateByKey 操作**：DStream API 为状态处理公开的另一个特殊操作是，一旦计算出状态就会使用新信息不断更新状态。以 Web 服务器日志为例，我们需要计算 Web 服务器所提供的所有 GET 或 POST 请求的运行计数。这种类型的功能可以通过 updateStateByKey 操作来实现。

（2）**输出操作**：输出操作是处理应用各种转换产生最终输出的操作。其可能只是在控制台上打印，在缓存或任何外部系统中持久化，如 NoSQL 数据库。输出操作类似于 RDD 定义的动作，并触发 DStream 上由用户定义的所有转换（同样类似于 RDD）。从 Spark 1.5.1 开始，DStream 支持以下输出操作。

- print()：这是开发人员调试作业时最常用的操作之一。它在运行流应用的驱动程序节点的控制台上打印 DStream 中每个批处理数据的前 10 个元素。

- saveAsTextFiles(prefix, suffix)：这会将 DStream 中的内容保留为文本文件。通过附加前缀和后缀生成每个批处理的文件名。

- saveAsObjectFiles(prefix, suffix)：这将 DStream 中的内容持久化为 Java 对象的序列文件。通过添加前缀和后缀来生成每个批处理的文件名。

- saveAsHadoopFiles(prefix, suffix)：这将 DStream 中的内容保留为 Hadoop 文件。通过添加前缀和后缀来生成每个批处理的文件名。

- foreachRDD(func)：这是处理输出时最重要、最广泛使用和最通用的函数之一。它将给定的函数 func 应用于从流生成的每个 RDD。此操作可编写自定义业务逻辑，以便在外部系统中持久化输出，如保存到 NoSQL 数据库或写入 Web 套接字。请注意，此功能由运行流应用的驱动程序节点来执行，这一点非常重要。

在本节，我们讨论了 Spark Streaming 的高层体系结构、组件和封装结构，还讨论了 DStream API 提供的各种转换和输出操作。现在继续向前，编写第一个 Spark Streaming 作业。

11.4 连接 Kafka 和 Spark Streaming

下面将完成一个程序，该程序将读取 Kafka 主题中的流数据并计算单词数。以下代码中将展示如下方面：Kafka 与 Spark Streaming 聚合；Spark 中 DStream 的创建和处理；查看流应用并读取无限无界的流以生成结果。

让我们来看一看下面的代码：

```java
package com.example.spark;
Import files:
import java.util.Collection;
import java.util.HashMap;
import java.util.Iterator;
import java.util.Map;
import java.util.regex.Pattern;

import org.apache.spark.SparkConf;
import org.apache.spark.api.java.function.Function;
import org.apache.spark.streaming.Duration;
import org.apache.spark.streaming.api.java.JavaDStream;
import org.apache.spark.streaming.api.java.JavaPairReceiverInputDStream;
import org.apache.spark.streaming.api.java.JavaStreamingContext;
import org.apache.spark.streaming.kafka.KafkaUtils;
import org.codehaus.jackson.map.DeserializationConfig.Feature;
import org.codehaus.jackson.map.ObjectMapper;
import org.codehaus.jackson.type.TypeReference;

import scala.Tuple2;

Then main classes:
public class JavaKafkaWordCount {
    private static final Pattern SPACE = Pattern.compile("");

    private JavaKafkaWordCount() {
    }

    @SuppressWarnings("serial")
    public static void main(String[] args) throws InterruptedException {
```

```
// if (args.length < 4) {
// System.err.println("Usage: JavaKafkaWordCount
<zkQuorum><group><topics><numThreads>");
// System.exit(1);
// }
```

定义数组。

```
args = new String[4];
    args[0]="localhost:2181";
    args[1]= "1";
    args[2]= "test";
    args[3]= "1";
```

定义方法。

```
    SparkConf sparkConf = new
SparkConf().setAppName("JavaKafkaWordCount").setMaster("spark://ImpetusNL163U:7077");
    // Create the context with a 1 second batch size
    JavaStreamingContext jssc = new JavaStreamingContext(sparkConf, new
Duration(20000));
```

参数转化。

```
    int numThreads = Integer.parseInt(args[3]);
    Map<String, Integer> topicMap = new HashMap<String, Integer>();
    String[] topics = args[2].split(",");
    for (String topic: topics) {
      topicMap.put(topic, numThreads);
    }
```

接收参数。

```
    JavaPairReceiverInputDStream<String, String> messages =
        KafkaUtils.createStream(jssc, args[0], args[1], topicMap);

    final JavaDStream<String> lines = messages.map(new
Function<Tuple2<String,String>, String>() {
@Override
public String call(Tuple2<String, String> v1) throws Exception {
ObjectMapper objectMapper = new ObjectMapper();
objectMapper.configure(Feature.USE_ANNOTATIONS, false);
Map<String,String> mapValue = objectMapper.readValue(v1._2(), new
TypeReference<Map<String,String>>() {
});
```

调整变量类型。

第 11 章 Spark Streaming

```java
Collection<String> values = mapValue.values();
String finalString = "";
for (Iterator<String> iterator = values.iterator(); iterator.hasNext();) {
String value = iterator.next();
if(finalString.length()==0){
finalString = finalString +value;
}else {
finalString = finalString+","+ value;
} }
```

带参数的返回函数。

```java
return finalString;
}
});
      lines.print();
      new Thread(){
public void run() {
while(true){
try {
Thread.sleep(1000);
} catch (InterruptedException e) {
// TODO Auto-generated catch block
e.printStackTrace();
}
System.out.println("###########################################################################"+lines.count());
}
};
    }.start();

    jssc.start();
    jssc.awaitTermination();
  }
}
```

11.5 小结

在本章，我们向读者介绍了 Apache Spark Streaming 的概念及其在 Spark 下的实现，研究了 Spark 和其他流处理平台之间的语义和差异，还让读者了解了在什么情况下 Spark 是比 Storm 更好的选择。我们通过介绍 API 和操作，然后给出与 Kafka 集成的字数统计示例，帮助读者理解了语言的语义。

第五部分
使用 Flink 实现实时分析

- 第 12 章　运用 Apache Flink

第 12 章
运用 Apache Flink

在本章，我们将向读者介绍实时处理的候选方案 Apache Flink。这里大多数数据源在跟踪和接收方面附件都保持不变，但在计算方法改为 Flink、集成和拓扑时，会非常不同。通过学习本章的内容，读者将理解并实现端到端的 Flink 过程，以便在实时流数据上解析、转换和聚合计算。

---- 本章主要包括以下内容 ----

- Flink 体系结构和执行引擎
- Flink 的基本组件和进程
- 将源流集成到 Flink
- Flink 处理和计算
- Flink 持久化
- Flink CEP
- Gelly
- 小试牛刀
- 源到 sink：Flink 的执行
- 在 Flink 上运行 Storm 拓扑

12.1 Flink 体系结构和执行引擎

Flink 是一个分布式流和批数据处理平台。Flink 的内核是一个流处理数据流引擎。Flink 基于 Kappa 架构——Kappa 架构是 Jay Kreps 于 2014 年提出的，该架构弥补了

Lambda 架构的缺陷。因此，在讨论 Flink 的细节之前，让我们先了解一下 Flink 的基础：Kappa 体系结构。Kappa 体系结构被设计使用单个数据流来实现实时数据处理和连续数据再处理。与 Lambda 架构有关的两个主要问题是：在同一源数据上维护两个不同的代码库，即实时分析和批分析代码库；事件的再处理需要更改代码，如图 12-1 所示，代码更改并不容易维护。

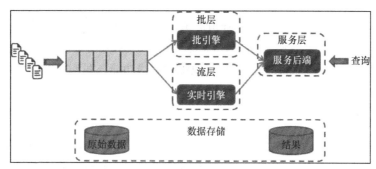

图 12-1

因此，在 Kappa 体系结构中，流数据包含所有内容。当发生故障和进行事件再处理时，将从再处理事件的 ETL 工具中开始一个新的进程，并使用另一个流将其反馈到服务层。事件的批具有可定义的开始和结束，并且流是无限的。换句话说，批是有界的，流是无界的。因此，批是流的一个子集，可以使用流处理进行处理，如图 12-2 所示。

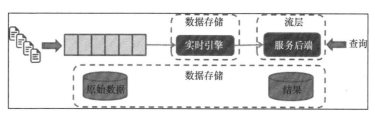

图 12-2

Flink 基于与 Kappa 架构相同的概念：使用单个流层进行实时和批处理。Flink 执行模型涵盖多种特点，如下所述。

- **"精确一次"语义**：Flink 在不增加处理负载的情况下定期维护检查点，从而确保了"精确一次"语义。
- **事件时间处理**：Flink 基于事件时间处理事件，这有助于维护事件的顺序，以及识别延迟传送的数据并对其采取适当的动作。

- **灵活的窗口**：具有支持持续时间、计数和基于会话的多用型窗口。窗口可以通过灵活的触发条件进行定制，以支持复杂的流模式。
- **高吞吐量和低延迟**：Flink 通过使用带屏障的检查点以及维护状态，提供高吞吐量和低延迟。
- **容错**：Flink 是一个轻量级的容错框架。Flink 中有检查点和状态快照这些概念，有助于实现完全容错。此外，源应该类似于可以回放和再处理的事件，例如 Apache Kafka。Flink 提供了与其他源的容错功能，我们将在本章的另一节中对此进行讨论。容错机制确保在发生故障时，不会处理任何事件两次，即"精确一次"处理。

容错机制以分布式流式数据流状态形式连续记录快照。此状态可根据配置存储在主节点或 HDFS 中。如果发生故障，Flink 将停止处理事件，系统重新启动操作算子并置位最后一个检查点，输入流将被置位为状态快照的点。下面将讨论什么是检查点以及如何维护状态。

基于 Chandy Lamport 算法在分布式流式数据流状态中拍摄快照，如图 12-3 所示。Flink 中的检查点基于两个概念：barriers 和状态。barriers 是 Flink 的核心元素，它随事件一起注入到分布式流式数据流状态中。barriers 将记录分组。每个 barriers 都有自己的唯一 ID。

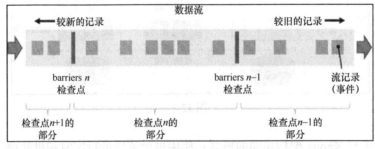

图 12-3

快照 n 的 barriers 被插入的位置（记为 S_n）是快照所包含的数据在数据源中的最大位置。此位置 S_n 将被报告给检查点协调器（Flink 的 **JobManager**）。

barriers 向下游流动。当中间算子从所有输入流中收到快照 n 的 barriers 时，它会将快照 n 的 barriers 发射到所有输出流中。一旦接收算子从所有输入流接收到屏障 n，它就会向检查点协调器确认快照 n 完成。在所有接收算子确认快照后，即视为已完成快照。

一旦快照 n 完成，作业将再也不会向数据源请求 S_n 之前的记录，因为此时这些记录已经通过整个数据流拓扑。

如果算子包含任何形式的状态，那么此状态也必须是快照的一部分。算子状态有不同的形式。

- **用户定义的状态**：这是由转换函数直接创建和修改的状态，如 `map()` 或 `filter()`。
- **系统状态**：此状态是指作为一部分算子计算的数据缓冲区。窗口缓冲区是此状态的典型示例，在该缓冲区中，系统收集（并聚合）窗口中的记录，直到对窗口进行计算并将其收回。

在从输入流接收到所有快照 barriers 后，并在向其输出流发出 barriers 之前，算子对自己的状态进行快照。此时，将根据 barriers 之前的记录对所有状态进行更新，而不会根据应用 barriers 后的记录进行任何更新。由于快照的状态可能很大，因此它存储在可配置的状态后端中。默认情况下，将会是 JobManager 内存，但在生产使用中，应配置分布式稳定存储（如 HDFS）。存储状态后，算子确认检查点，将快照屏障发送到输出流中，然后继续操作。

12.2 Flink 的基本组件和进程

Flink 的一些流程和基本组件如下。

Flink 中有两种类型的**进程**，如图 12-4 所示。

- **JobManager(Master)**：它负责在 **TaskManagers** 中分配任务，维护检查点，并在出现任何故障时进行恢复。为了实现高可用性，需要设置多个 **JobManager**，其中一个作为 leader，另一个作为备用。leader 和备用 JobManager 始终保持同步，一旦 leader 发生故障，备用 JobManager 将成为 leader。
- **TaskManager (worker)**：TaskManager 负责执行任务、维护算子的缓冲区等。TaskManager 是一个 JVM 进程，子任务在称为任务槽的多个线程中执行。TaskManager 可以有多个槽，一个槽可以由作业的多个子任务共享。

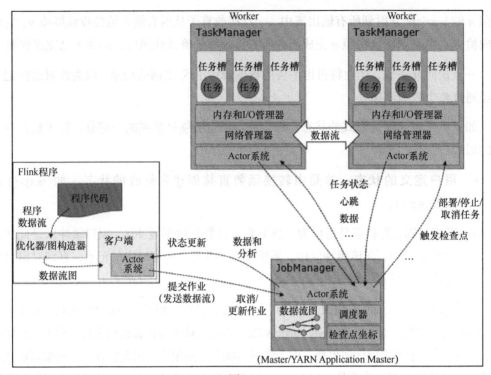

图 12-4

客户端不是程序执行的一部分。它只是提交作业,可能会保持连接,以定期获得作业的状态。

组件:Flink 是具有不同堆栈层的分层系统,如图 12-5 所示。

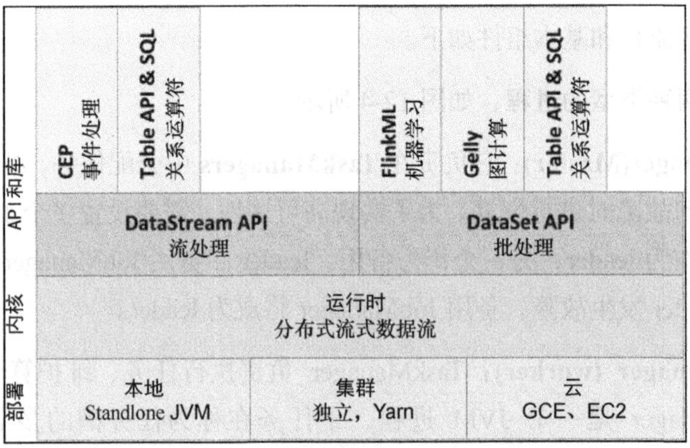

图 12-5

- JobGraph 是由 **Datastream** 和 **DataSet** API 通过单独编译生成的程序。**Datastream API** 用于分布式数据流的实时处理，**DataSet API** 用于分布式数据流的批处理。
- 运行时层接收 JobGraph 形式的程序，JobGraph 在各种可用的部署选项上执行它们，如本地、集群或云。

12.3 将源流集成到 Flink

可以使用多个源进行集成，如 Apache Kafka、Amazon Kinesis 流、RabbitMQ、Apache NiFi、Twitter Streaming API。

接下来将看到与 Apache Kafka 和 RabbitMQ 集成的演示。

12.3.1 和 Apache Kafka 集成

在前面的章节中，我们已经讨论了 Apache Kafka 的设置，因此这里将重点介绍集成 Flink 和 Kafka 的 Java 代码。

按照下面给定的步骤进行操作。

① 在 `pom.xml` 中添加依赖项，代码如下：

```xml
<dependency>
    <groupId>org.apache.flink</groupId>
    <artifactId>flink-streaming-java_2.11</artifactId>
    <version>1.2.0</version>
</dependency>
```

所有类型的集成都需要以上的依赖项。以下是 Flink 和 Kafka 集成所需的特定依赖关系。

```xml
<dependency>
    <groupId>org.apache.flink</groupId>
    <artifactId>flink-connector-kafka-0.8_2.11</artifactId>
    <version>1.2.0</version>
</dependency>
<dependency>
    <groupId>org.apache.kafka</groupId>
    <artifactId>kafka_2.11</artifactId>
    <version>0.8.2.2</version>
</dependency>
```

② 设置 Kafka 源，代码如下：

```
Properties properties = new Properties();
properties.setProperty("bootstrap.servers", "localhost:9092");
properties.setProperty("zookeeper.connect", "localhost:2181");
properties.setProperty("group.id", "test");
properties.setProperty("auto.offset.reset", "latest");
FlinkKafkaConsumer08<String> flinkKafkaConsumer08 = new
FlinkKafkaConsumer08<>("flink-test", new SimpleStringSchema(),
properties);
```

创建包含与 Kafka 相关信息的属性，如 Broker、Zookeeper、Grroup id 和 offset。Flink 软件包中 Kafka 0.8 的 `FlinkKafkaConsumer08` 可用，由此创建一个以参数为主题名和消息序列化器的对象。

③ 创建 Streaming 环境，代码如下：

```
StreamExecutionEnvironment env =
StreamExecutionEnvironment.getExecutionEnvironment();
```

现在，在 Flink 中创建 Streaming 环境的一个对象。

④ 在 Streaming 环境中添加 Kafka 源，代码如下：

```
DataStream<String> messageStream =
env.addSource(flinkKafkaConsumer08);
```

当 Kafka 源添加到 Flink 环境中时，它将返回数据类型为 `DataStream` 的对象。

⑤ 读取源并在控制台上打印，代码如下：

```
messageStream.rebalance().map(new MapFunction<String, String>() {
    private static final long serialVersionUID =
-6867736771747690202L;
    @Override
    public String map(String value) throws Exception {
        return "Kafka and Flink says: " + value;
    }
}).print();
```

在消息流上应用 `map()` 函数来读取消息。

⑥ 执行环境，代码如下：

```
env.execute();
```

读者必须执行环境，否则程序将不会在 Flink 上执行。这相当于创建 DAG，而不是在 JobManager 上提交。

示例

Kafka 需要以下输入，输出将显示在控制台上。

输入：打开控制台，并使用命令行添加关于 Kafka 主题的消息，如图 12-6 所示。

图 12-6

输出：在配套资源中转到 `FlinkKafkaSourceExexample`，右键单击执行程序，选择 **Run as Java Application** 选项。输出将显示在控制台上，如图 12-7 所示。

```
Connected to JobManager at Actor[akka://flink/user/jobmanager_1#-1479811007]
06/27/2017 19:51:35    Job execution switched to status RUNNING.
06/27/2017 19:51:35    Source: Custom Source(1/4) switched to SCHEDULED
06/27/2017 19:51:35    Source: Custom Source(1/4) switched to DEPLOYING
06/27/2017 19:51:35    Source: Custom Source(2/4) switched to SCHEDULED
06/27/2017 19:51:35    Source: Custom Source(2/4) switched to DEPLOYING
06/27/2017 19:51:35    Source: Custom Source(3/4) switched to SCHEDULED
06/27/2017 19:51:35    Source: Custom Source(3/4) switched to DEPLOYING
06/27/2017 19:51:35    Source: Custom Source(4/4) switched to SCHEDULED
06/27/2017 19:51:35    Source: Custom Source(4/4) switched to DEPLOYING
06/27/2017 19:51:35    Map -> Sink: Unnamed(1/4) switched to SCHEDULED
06/27/2017 19:51:35    Map -> Sink: Unnamed(1/4) switched to DEPLOYING
06/27/2017 19:51:35    Map -> Sink: Unnamed(2/4) switched to SCHEDULED
06/27/2017 19:51:35    Map -> Sink: Unnamed(2/4) switched to DEPLOYING
06/27/2017 19:51:35    Map -> Sink: Unnamed(3/4) switched to SCHEDULED
06/27/2017 19:51:35    Map -> Sink: Unnamed(3/4) switched to DEPLOYING
06/27/2017 19:51:35    Map -> Sink: Unnamed(4/4) switched to SCHEDULED
06/27/2017 19:51:35    Map -> Sink: Unnamed(4/4) switched to DEPLOYING
06/27/2017 19:51:36    Source: Custom Source(2/4) switched to RUNNING
06/27/2017 19:51:36    Source: Custom Source(4/4) switched to RUNNING
06/27/2017 19:51:36    Source: Custom Source(1/4) switched to RUNNING
06/27/2017 19:51:36    Source: Custom Source(3/4) switched to RUNNING
06/27/2017 19:51:36    Map -> Sink: Unnamed(1/4) switched to RUNNING
06/27/2017 19:51:36    Map -> Sink: Unnamed(2/4) switched to RUNNING
06/27/2017 19:51:36    Map -> Sink: Unnamed(3/4) switched to RUNNING
06/27/2017 19:51:36    Map -> Sink: Unnamed(4/4) switched to RUNNING
1> Kafka and Flink says: hi
2> Kafka and Flink says: this
3> Kafka and Flink says: is
1> Kafka and Flink says: and
4> Kafka and Flink says: flink
2> Kafka and Flink says: kafka
3> Kafka and Flink says: integration
4> Kafka and Flink says: example
```

图 12-7

12.3.2 和 RabbitMQ 集成

下载 RabbitMQ 并进行配置。要将 Flink 与 RabbitMQ 集成，需要执行以下步骤。

① 在 pom.xml 中添加依赖项，代码如下：

```xml
<dependency>
    <groupId>org.apache.flink</groupId>
    <artifactId>flink-connector-rabbitmq_2.11</artifactId>
    <version>1.2.0</version>
</dependency>
```

② 设置 RMQ Publisher，代码如下：

```
ConnectionFactory factory = new ConnectionFactory(); // line 1
factory.setUsername("guest");
factory.setPassword("guest");
factory.setVirtualHost("/");
factory.setHost("localhost");
factory.setPort(5672);
Connection newConnection = factory.newConnection(); // line 2
Channel channel = newConnection.createChannel(); // line 3
Scanner scanner = new Scanner(System.in); // line 5
String message = "";
while(!message.equals("exit")){ // line 6
    System.out.println("Enter your message");
    message = scanner.next();
    channel.queueDeclare("flink-test", true, false, false, null);
// line 7
    channel.basicPublish("", "flink-test", new
BasicProperties.Builder()
.correlationId(java.util.UUID.randomUUID().toString()).build();,
    message.getBytes()); // line 8
}
```

要在 Flink 中使用 RabbitMQ 进行消息的"精确"一次处理，需要给每条消息添加独一无二的 correlationId。在前面的程序中，我们在第 1 行使用 RabbitMQ 服务器创建了 connection factory。在第 2 行，从 connection factory 处获取新连接。在第 3 行通过连接获取通道。在第 5 行从控制台的用户上获取消息并循环请求，直到消息不存在于第 6 行中退出。在第 7 行使用所需的信息在通道中声明队列。在第 8 行，队列 flink-test 发布消息，并在基本属性中设置唯一的 correlationId。

③ 设置 RMQ 源，代码如下：

```
final RMQConnectionConfig connectionConfig = new
RMQConnectionConfig.Builder.setHost("localhost").setPort(5672).setV
irtualHost("/").setUserName("guest").setPassword("guest").build();
```

现在，Flink 程序从包含 RMQ 服务器所需详细信息的 RMQ 连接配置开始。

④ 创建 Streaming 环境，代码如下：

```
final StreamExecutionEnvironment env =
StreamExecutionEnvironment.getExecutionEnvironment();
env.enableCheckpointing(30000, CheckpointingMode.EXACTLY_ONCE);
```

创建 Streaming 环境并启用检查点，将时间和策略设置为 EXACTLY_ONCE。如果想使用 RMQ 和 Flink "精确一次"语义，这是必需的设置。否则为至少一次语义。

⑤ 在 Streaming 环境中添加 RMQ 源，代码如下：

```
final DataStream<String> stream = env.addSource(new
RMQSource<String>(connectionConfig,"flink-test", true, new
SimpleStringSchema())).setParallelism(1);
```

在 Streaming 环境中添加 RabbitMQ 源。RMQ 源接收参数，如队列名、是否设置 correlationId 以及消息反序列化器。

⑥ 读取源并在控制台上打印，代码如下：

```
messageStream.rebalance().map(new MapFunction<String, String>() {
    private static final long serialVersionUID =
-6867736771747690202L;
    @Override
    public String map(String value) throws Exception {
        return "RabbitMQ and Flink says: " + value;
    }
}).print();
```

⑦ 执行环境，代码如下：

```
env.execute();
```

运行示例

要使用 RMQ Publisher，则在 RMQ 上提供以下输入是必需的，并且输出将显示在控制台上。

输入：在 Eclipse 中运行 RMQ Publisher 发行者程序。在控制台的 RMQ 中输入消息，如图 12-8 所示。

```
Enter your message
hi
Enter your message
this
Enter your message
is
Enter your message
RMQ
Enter your message
and
Enter your message
Flink
Enter your message
integration
Enter your message
example
Enter your message
```

图 12-8

RMQ UI 显示队列中的 8 条消息，如图 12-9 所示。

图 12-9

输出：在 Eclipse 中运行 `FlinkRabbitMQSourceExample` 程序，在读取 RMQ 队列中的所有消息后，输出将显示在控制台上，此处不能保证消息的顺序。输出如图 12-10 所示。

```
Connected to JobManager at Actor[akka://flink/user/jobmanager_1#-1293803841]
06/28/2017 18:26:29    Job execution switched to status RUNNING.
06/28/2017 18:26:29    Source: Custom Source(1/1) switched to SCHEDULED
06/28/2017 18:26:29    Source: Custom Source(1/1) switched to DEPLOYING
06/28/2017 18:26:29    Map -> Sink: Unnamed(1/4) switched to SCHEDULED
06/28/2017 18:26:29    Map -> Sink: Unnamed(1/4) switched to DEPLOYING
06/28/2017 18:26:29    Map -> Sink: Unnamed(2/4) switched to SCHEDULED
06/28/2017 18:26:29    Map -> Sink: Unnamed(2/4) switched to DEPLOYING
06/28/2017 18:26:29    Map -> Sink: Unnamed(3/4) switched to SCHEDULED
06/28/2017 18:26:29    Map -> Sink: Unnamed(3/4) switched to DEPLOYING
06/28/2017 18:26:29    Map -> Sink: Unnamed(4/4) switched to SCHEDULED
06/28/2017 18:26:29    Map -> Sink: Unnamed(4/4) switched to DEPLOYING
06/28/2017 18:26:29    Source: Custom Source(1/1) switched to RUNNING
06/28/2017 18:26:29    Map -> Sink: Unnamed(1/4) switched to RUNNING
06/28/2017 18:26:29    Map -> Sink: Unnamed(2/4) switched to RUNNING
06/28/2017 18:26:29    Map -> Sink: Unnamed(3/4) switched to RUNNING
06/28/2017 18:26:29    Map -> Sink: Unnamed(4/4) switched to RUNNING
4> RabbitMQ and Flink says: RMQ
1> RabbitMQ and Flink says: hi
4> RabbitMQ and Flink says: example
2> RabbitMQ and Flink says: this
1> RabbitMQ and Flink says: and
2> RabbitMQ and Flink says: Flink
3> RabbitMQ and Flink says: is
3> RabbitMQ and Flink says: integration
```

图 12-10

12.4　Flink 处理和计算

我们在前面讨论了 Flink 不同的源，现在将讨论源数据的处理。根据数据源的类型（有界或无界），Flink 中提供了处理 API。Datastream API 可用于无界数据源，DataSet API 可用于有界数据源。

12.4.1　Datastream API

在转到转换函数之前，让我们看一下 Datastream 的一个示例。以下代码段是使用 Datastream API 实现的字数统计示例：

```
StreamExecutionEnvironment env =
StreamExecutionEnvironment.getExecutionEnvironment();
DataStream<Tuple2<String, Integer>> dataStream = env
.socketTextStream("localhost", 9999)
.flatMap(new Splitter())
.keyBy(0)
.timeWindow(Time.seconds(5))
.sum(1);
```

`StreamExecutionEnvironment` 用来创建 DataStream 对象。在前面的示例中，环境首先使用套接字连接数据源。然后，应用了 `flatMap` 函数，该函数将句子分成单词的元组，并将其计数为 1。KeyBy 根据键（即字）对流进行分组。应用一个 5s 的 `timeWindow` 函数来执行字数计算。最后，应用求和的聚合函数将所有字的计数相加。

以下是 Datastream API 提供的不同转换。

- `map`：它是最简单的转换。输入/输出参数的数量没有变化，只是改变/修改了值或改变了值的类型。

```
dataStream.map(new MapFunction<Integer, Integer>() {
    @Override
    public Integer map(Integer value) throws Exception {
        return value + 2;
    }
});
```

- `FlatMap`：FlatMap 返回集合中的值，它可能返回一个或多个值或不返回值。

```
dataStream.flatMap(new FlatMapFunction<String, String>() {
```

```
@Override
    public void flatMap(String value, Collector<String> out)
    throws Exception {
      for(String word: value.split(" ")){
       out.collect(word);
      }
    }
});
```

- Filter：Filter 转换用于筛选数据流中的事件。

```
dataStream.filter(new FilterFunction<Integer>() {
    @Override
    public boolean filter(Integer value) throws Exception {
        return value != 0;
    }
});
```

- KeyBy：它对流进行分区，并且每个分区都具有相同键的事件，并返回 keyed Stream。如果没有重写 hashCode 方法，那么键不能是数组和类对象。

```
dataStream.keyBy(0)
```

- reduce：它对当前值和最后一个记录值执行指定的动作，并生成一个新值。KeyBy 和 reduce 之间的区别在于，KeyBy 在分区上工作，因此重排较少，但 reduce 对流中需要重排的每个事件都执行一次函数。

```
keyedStream.reduce(new ReduceFunction<Integer>() {
@Override
    public Integer reduce(Integer value1, Integer value2)
    throws Exception {
        return value1 * value2;
    }
});
```

- fold：fold 与 reduce 相同，唯一的区别是 fold 在执行指定的函数之前指定一个种子值。示例如下：

```
DataStream<String> result =
keyedStream.fold(1, new FoldFunction<Integer, Integer>() {
@Override
    public Integer fold(Integer current, Integer value) {
        return current * value;
    }
});
```

- `Aggregations`：有多个可用聚合函数，示例如下：

```
keyedStream.sum(0);
keyedStream.sum("key");
keyedStream.min(0);
keyedStream.min("key");
keyedStream.max(0);
keyedStream.max("key");
keyedStream.minBy(0);
keyedStream.minBy("key");
keyedStream.maxBy(0);
keyedStream.maxBy("key");
```

- `Window`：`Window` 可以在已分区的 `keyedStream` 上定义。窗口根据用户指定的时间对每个键中的数据进行分组。示例如下：

```
dataStream.keyBy(0).timeWindow(Time.seconds(5))
```

12.4.2 DataSet API

DataSet API 用于批处理。它具有与 Datastream API 几乎相同的类型转换。以下代码段是使用 DataSet API 进行字数统计的一个小示例。

```
final ExecutionEnvironment env =
ExecutionEnvironment.getExecutionEnvironment();
DataSet<String> text = env.fromElements(
"Who's there?",
"I think I hear them. Stand, ho! Who's there?");
DataSet<Tuple2<String, Integer>> wordCounts = text
.flatMap(new LineSplitter())
.groupBy(0)
.sum(1);
```

这里的执行环境与 Datastream 不同，即为 `ExecutionEnvironment`。上面的程序正在执行相同的任务，但对有界数据使用不同的方法。

以下是 DataSet API 提供的转换，它们不同于 Datastream API。

- `distinct()`：返回数据集中互不相同的元素。示例如下：

```
data.distinct();
```

- `join()`：转换基于键组合两个数据集。示例如下：

```
result = input1.join(input2)
.where(0) // key of the first input (tuple field 0)
```

```
.equalTo(1); // key of the second input (tuple field 1)
```

- union：两个数据集的并集。示例如下：

```
DataSet<String> result = data1.union(data2);
```

- First-n：返回数据集中前 *n* 个元素。示例如下：

```
DataSet<Tuple2<String,Integer>> result1 = in.first(3);
```

我们已经讨论了源、处理和计算，接下来讨论 Flink 支持的 sink。

12.5　Flink 持久化

Flink 提供具有 sink 或持久性的连接器，例如 Apache Kafka、Elasticsearch、Hadoop Filesystem、RabbitMQ、Amazon Kinesis Streams、Apache NiFi、Apache Cassandra。

接下来将讨论 Flink 和 Cassandra 的连接，因为它是最流行的。

和 Cassandra 集成

在前面的章节中，我们讨论和解释了 Cassandra 的设置，因此现在将直接介绍在 Flink 和 Cassandra 之间建立连接所需的程序。

- 在 pom.xml 中添加依赖项。

```xml
<dependency>
    <groupId>org.apache.flink</groupId>
    <artifactId>flink-connector-cassandra_2.11</artifactId>
    <version>1.2.0</version>
</dependency>
<dependency>
    <groupId>com.codahale.metrics</groupId>
    <artifactId>metrics-json</artifactId>
    <version>3.0.2</version>
</dependency>
```

- 创建数据流。

```
 DataStream<Tuple4<Long,Integer,Integer,Long>> messageStream =
env.addSource(flinkKafkaConsumer08).map(new MapFunction<String,
Tuple4<Long,Integer,Integer,Long>>() {
    private static final long serialVersionUID =
4723214570372887208L;
```

```
        @Override
            public Tuple4<Long,Integer,Integer,Long> map(String input) 
throws          Exception
            {
                String[] inputSplits = input.split(",");
                return Tuple4.of(Long.parseLong(inputSplits[0]),
                Integer.parseInt(inputSplits[1]),
                Integer.parseInt(inputSplits[2]),
                Long.parseLong(inputSplits[3]));
            }
});
```

要与 Cassandra sink 集成,必须将字符串中的数据流转换为元组。因为有 4 列需要在 Cassandra 中持久化,所以对所需的数据类型使用了 `Tuple4`。应用 `map` 函数将值从一种格式转换为另一种格式。

- 创建 `CassandraSink`。

```
CassandraSink.addSink(messageStream).setQuery("INSERT INTO 
tdr.packet_tdr (phone_number, bin, bout, timestamp) values (?, ?, ?
,?);").setClusterBuilder(new ClusterBuilder() {
   private static final long serialVersionUID = 1L;
   @Override
      public Cluster buildCluster(Cluster.Builder builder) {
         return builder.addContactPoint("127.0.0.1").build();
      }
}).build();
```

`CassandraSink` 用静态方法来添加输入数据流 `addSink`。可以使用 `setQuery` 方法添加查询。`setClusterBuilder` 用于定义 Cassandra 集群的详细信息。

运行示例

要运行该示例,必须运行数据生成器,`DataGenerator` 将数据推送到 Kafka 主题,然后是 `FlinkCassandraConnector` 的 Flink 程序从 Kafka 主题中读取数据并将其推送到 Cassandra。

输入:在 Eclipse 中运行该程序,在控制台上显示图 12-11 所示的内容。数据集将是不同的,因为它是随机的。

输出:运行 `FlinkCassandraConnector` 后,控制台上不会打印任何内容,因此必须检查 Cassandra 中是否存在数据。运行图 12-12 所示的查询语句。

```
9999999983,7086759,8629334,1499350308660
9999999953,3927537,5309044,1499350308926
9999999993,2433812,7203793,1499350308926
9999999976,7046874,7056789,1499350308927
9999999997,9897191,4827991,1499350308928
9999999975,6247369,6145325,1499350308933
9999999961,1320813,4681550,1499350308934
9999999973,6989620,7262999,1499350308934
9999999971,2767230,8894962,1499350308934
9999999984,6499812,4954905,1499350308934
9999999968,5716208,6606575,1499350308934
9999999990,7270925,9005006,1499350308935
9999999995,3363460,1700805,1499350308935
9999999986,5162460,9524685,1499350308935
9999999951,3616511,3256483,1499350308935
```

图 12-11

图 12-12

12.6　Flink CEP

CEP 代表复杂事件处理。Flink 提供了在高吞吐量和低延迟的数据流上实施 CEP 的 API。CEP 是一种处理数据流，它将规则或条件予以应用，任何满足条件的事件都将保存在数据库中，并向用户发送通知。Flink CEP 的运行流程如图 12-13 所示。Flink 针对流中的每个事件匹配复杂模式。此过程过滤出有用的事件，并丢弃不相关的事件。由此使用户有机会快速掌握数据中真正重要的内容。举个智能发电机组的例子，该例可以发送发电机状态和系统温度。假设发电机组的温度超过 40℃，用户应该收到一条通知，让它在一段时间内关闭，或者立即采取行动，以免发生事故。

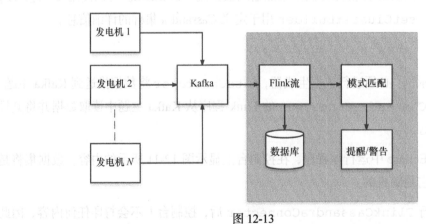

图 12-13

12.7 Pattern API

Apache Flink 提供了在数据流上应用复杂事件处理的 Pattern API。下面展示了一些重要的方法。

- `begin`：定义了模式的启动状态，如下所示：

```
Pattern<Event, ?> start = Pattern.<Event>begin("start");
```

- `followedBy`：会附加一个新的模式状态。这里，在两个匹配事件之间可能会发生其他事件，如下所示：

```
Pattern<Event, ?> followedBy = start.followedBy("next");
```

- `where`：为当前模式状态定义了一个筛选条件，如果事件通过了筛选，那么它可以匹配该状态，如下所示：

```
patternState.where(new FilterFunction <Event>() {
@Override
    public boolean filter(Event value) throws Exception {
        return ... // some condition
    }
});
```

- `within`：定义了事件序列与被丢弃的 pattern 的最大时间间隔，如下所示：

```
patternState.within(Time.seconds(10));
```

- `subtype(subClass)`：这将为当前模式定义子类型条件。只有在事件属于此子类型的情况下，事件才能与模式匹配，如下所示：

```
pattern.subtype(SubEvent.class);
```

12.7.1 检测模式

一旦创建了一个模式，就必须在数据流上应用该模式进行匹配。以下是通过代码执行此操作的方法。

```
DataStream<Event> input = ...
Pattern<Event, ?> pattern = ...
PatternStream<Event> patternStream = CEP.pattern(input, pattern);
```

在这里，必须使用创建的输入数据流和模式来创建 `PatternStream`。现在，`PatternStream` 将拥有与所定义模式匹配的事件。

12.7.2 模式选择

既然已经创建了模式，并将其应用于数据流，那么如何获得匹配的事件？这可以通过使用以下代码段来实现。

```
class MyPatternSelectFunction<IN, OUT> implements PatternSelectFunction<IN, OUT> {
@Override
    public OUT select(Map<String, List<IN>> pattern) {
        IN startEvent = pattern.get("start").get(0);
        IN endEvent = pattern.get("end").get(0);
        return new OUT(startEvent, endEvent);
    }
}
```

`In` 是应用事件模式的类，作为匹配事件的动作 `Out` 类是输出形式。

12.7.3 示例

下面的代码解释了可用的模式 API 及其使用方式。在此示例中，如果移动/物联网设备在 10s 内生成超过 15MB 的数据，则将会发出警告。

```
Pattern<DeviceEvent, ?> alertPattern =
Pattern.<DeviceEvent>begin("first").subtype(DeviceEvent.class).where(new DeviceFilterFunction()).followedBy("second").subtype(DeviceEvent.class).where(new DeviceFilterFunction()).within(Time.seconds(TXN_TIMESPAN_SEC));
```

前面的代码片段已经定义了与数据流的每个事件相匹配的模式。

```
PatternStream<DeviceEvent> tempPatternStream =
CEP.pattern(messageStream.rebalance().keyBy("phoneNumber"), alertPattern);
```

前面的代码片段将模式应用到数据流上，得到只包含匹配事件的模式流。

```
DataStream<DeviceAlert> alert = tempPatternStream.select(new
PatternSelectFunction<DeviceEvent, DeviceAlert>() {
    private static final long serialVersionUID = 1L;
    @Override
        public DeviceAlert select(Map<String, DeviceEvent> pattern) {
            DeviceEvent first = (DeviceEvent) pattern.get("first");
```

```
                DeviceEvent second = (DeviceEvent) pattern.get("second");
                allTxn.clear();
                allTxn.add(first.getPhoneNumber() + " used " + ((first.getBin()
                + first.getBout())/1024/1024) +" MB at " + new
                Date(first.getTimestamp()));
                allTxn.add(second.getPhoneNumber() + " used " +
                ((second.getBin() + second.getBout())/1024/1024) +" MB at " +
                new Date(second.getTimestamp()));
                return new DeviceAlert(first.getPhoneNumber(), allTxn);
            }
});
```

在此之前，我们可以使用 `select` 函数对匹配事件执行操作。完整的代码可以在配套资源中找到。要运行该示例，首先启动在 Kafka 上推送数据的 `DataGenerator`，然后启动 `DeviceUsageMonitoring`。

12.8 Gelly

Gelly 是 Flink 的图 API。在 Gelly 中，可以创建、转换和修改图。Gelly API 提供了图分析中所有的基本和高级功能，也可以选择不同的图算法。

Gelly API

Gelly 为 API 提供了对图应用动作的能力，我们将在后面的章节中讨论这些 API。

1. 图表示

图由顶点和边组成的数据集表示。图节点由 `Vertex` 类型表示，顶点由唯一 ID 和值定义。可将没有值的顶点定义为 `NullValue`。以下是在图中创建顶点的方法。

```
Vertex<String, Long> v = new Vertex<String, Long>("vertex 1", 8L);
Vertex<String, NullValue> v = new Vertex<String, NullValue>("vertex 1",
NullValue.getInstance());
```

图的边由边类型表示。边由源 ID（源顶点的 ID）、目标 ID（目标顶点的 ID）和可选值定义。源 ID 和目标 ID 的类型应与顶点 ID 的相同。以下是在图中创建 `Edge` 的方法。

```
Edge<String, Double> e = new Edge<String, Double>("vertex 1", "vertex 2",0.5 );
```

2. 图创建

可以根据 `ExecutionEnvironment` 中的以下语句创建图。

- 利用文本文件创建图。

```
ExecutionEnvironment env =
ExecutionEnvironment.getExecutionEnvironment();
DataSet<Tuple2<String, Long>> vertexTuples =
env.readCsvFile("path/to/vertex/input").types(String.class,
Long.class);
DataSet<Tuple3<String, String, Double>> edgeTuples =
env.readCsvFile("path/to/edge/input").types(String.class,
String.class, Double.class);
Graph<String, Long, Double> graph =
Graph.fromTupleDataSet(vertexTuples, edgeTuples, env);
```

- 利用集合创建图。

```
ExecutionEnvironment env =
ExecutionEnvironment.getExecutionEnvironment();
List<Vertex<Long, Long>> vertexList = new ArrayList...
List<Edge<Long, String>> edgeList = new ArrayList...
Graph<Long, Long, String> graph = Graph.fromCollection(vertexList,
edgeList, env);
```

3. 图转换

以下是一些重要的图转换。

- `map`：该变换可以应用于顶点或边。顶点和边的 ID 保持不变，可以根据用户定义的函数更改该值。

```
Graph<Long, Long, Long> updatedGraph = graph.mapVertices(new
MapFunction<Vertex<Long, Long>, Long>() {
public Long map(Vertex<Long, Long> value) {
return value.getValue() + 1;
}
});
```

- `filter`：该变换可以过滤图中的顶点或边。如果在边上应用了过滤转换，则过滤后的边将保留在图中，顶点也不会被移除。如果对顶点应用了过滤转换，则过滤后的顶点将保留在图中。

```
graph.subgraph(new FilterFunction<Vertex<Long, Long>>() {
```

```
public boolean filter(Vertex<Long, Long> vertex) {
return (vertex.getValue() > 0);
}
},
new FilterFunction<Edge<Long, Long>>() {
public boolean filter(Edge<Long, Long> edge) {
return (edge.getValue() < 0);
}
})
```

- `reverse`：该方法可以返回一个新的图，其中所有边的方向都已反转。
- **Union**：在并集操作中，会删除重复的顶点，但会保留重复的边。

12.9 小试牛刀

我们在前面几节中展示了许多示例，现在准备进行一些实际操作。所涉及的代码在配套资源中给出，以供读者参考。阅读配套资源中的 `README.MD` 可以了解程序的执行情况。

1. 源到 sink——Flink 的执行

读者已经看到了将 Apache Kafka 或 RabbitMQ 作为源与 Cassandra 作为 sink 进行集成的示例。现在，我们将集成 Apache Kafka 和 Elasticsearch2.x，其中 Apache Kafka 作为源而 Elasticsearch2.x 作为 sink。连接器将被更改，因为使用了 Elasticsearch 而不是 Cassandra。

首先导入文件。

```
package com.boof.flink.diy;
import java.net.InetAddress;
import java.net.InetSocketAddress;
import java.util.ArrayList;import java.util.HashMap;
import java.util.List;
import java.util.Map;
import java.util.Properties;
import org.apache.flink.api.common.functions.RuntimeContext;
import org.apache.flink.streaming.api.datastream.DataStream;
import
org.apache.flink.streaming.api.environment.StreamExecutionEnvironment;
import
org.apache.flink.streaming.connectors.elasticsearch2.ElasticsearchSink;
import
org.apache.flink.streaming.connectors.elasticsearch2.ElasticsearchSinkFunction;
```

```
import org.apache.flink.streaming.connectors.elasticsearch2.RequestIndexer;
import org.apache.flink.streaming.connectors.kafka.FlinkKafkaConsumer08;
import org.elasticsearch.action.index.IndexRequest;
import org.elasticsearch.client.Requests;
import com.book.flinkcep.example.DeviceEvent;
import com.book.flinkcep.example.DeviceSchema;
```

2. Flink Elasticsearch 集成主类

主类如下：

```
public class FlinkESConnector {
    public static void main(String[] args) throws Exception {
    Properties properties = new Properties();
    properties.setProperty("bootstrap.servers", "localhost:9092");
    properties.setProperty("zookeeper.connect", "localhost:2181");
    properties.setProperty("group.id", "test");
    properties.setProperty("auto.offset.reset", "latest");
    FlinkKafkaConsumer08<DeviceEvent> flinkKafkaConsumer08 = new
    FlinkKafkaConsumer08<>("device-data",
            new DeviceSchema(), properties);
        StreamExecutionEnvironment env =
        StreamExecutionEnvironment.getExecutionEnvironment();
        DataStream<DeviceEvent> messageStream =
        env.addSource(flinkKafkaConsumer08);
```

接下来，调用以下方法。

```
Map<String, String> config = new HashMap<>();
config.put("cluster.name", "my-application");
// This instructs the sink to emit after every element, otherwise they
would be buffered
config.put("bulk.flush.max.actions", "1");
        List<InetSocketAddress> transportAddresses = new ArrayList<>();
        transportAddresses.add(new
        InetSocketAddress(InetAddress.getByName("127.0.0.1"), 9300));
        messageStream.addSink(new
ElasticsearchSink<DeviceEvent>(config,
        transportAddresses, new ESSink()));
        env.execute(); }}
```

3. Elasticsearch sink 作为 Flink 的操作算子

接下来，将显示 JSON 对象的代码。

```java
class ESSink implements ElasticsearchSinkFunction<DeviceEvent> {
    private static final long serialVersionUID = -4286031843082751966L;
    @Override
    public void process(DeviceEvent element, RuntimeContext ctx,
RequestIndexer indexer) {
        Map<String, Object> json = new HashMap<>();
            json.put("phoneNumber", element.getPhoneNumber());
            json.put("bin", element.getBin());
            json.put("bout", element.getBout());
            json.put("timestamp", element.getTimestamp());
            System.out.println(json);
        IndexRequest source = Requests.indexRequest()
            .index("flink-test")
            .type("flink-log")
            .source(json);
        indexer.add(source);
    }
}
```

4. 保存事件数据的域类

下一个代码片段调用了 public 方法。

```java
 package com.book.flinkcep.example;

public class DeviceEvent {

    private long phoneNumber;
    private int bin;
    private int bout;
    private long timestamp;

    public long getPhoneNumber() {
        return phoneNumber;
    }

    public void setPhoneNumber(long phoneNumber) {
        this.phoneNumber = phoneNumber;
    }

    public int getBin() {
        return bin;
    }

    public void setBin(int bin) {
        this.bin = bin;
    }

    public int getBout() {
```

```java
            return bout;
        }

        public void setBout(int bout) {
            this.bout = bout;
        }

        public long getTimestamp() {
            return timestamp;
        }

        public void setTimestamp(long timestamp) {
            this.timestamp = timestamp;
        }
   public static DeviceEvent fromString(String line) {

            String[] tokens = line.split(",");

            if (tokens.length != 4) {
                throw new RuntimeException("Invalid record: " + line);
            }

            DeviceEvent deviceEvent = new DeviceEvent();

            deviceEvent.phoneNumber = Long.parseLong(tokens[0]);
            deviceEvent.bin = Integer.parseInt(tokens[1]);
            deviceEvent.bout = Integer.parseInt(tokens[2]);
            deviceEvent.timestamp = Long.parseLong(tokens[3]);

            return deviceEvent;
        }
        @Override
        public int hashCode() {
            final int prime = 31;
            int result = 1;
            result = prime * result + bin;
            result = prime * result + bout;
            result = prime * result + (int) (phoneNumber ^ (phoneNumber >>>32));
            result = prime * result + (int) (timestamp ^ (timestamp >>>32));
            return result;
        }

        @Override
        public boolean equals(Object obj) {
            if (this == obj)
                return true;
            if (obj == null)
                return false;
            if (getClass() != obj.getClass())
```

```
                return false;
            DeviceEvent other = (DeviceEvent) obj;
            if (bin != other.bin)
                return false;
            if (bout != other.bout)
                return false;
            if (phoneNumber != other.phoneNumber)
                return false;
            if (timestamp != other.timestamp)
                return false;
            return true;
    }
    @Override
        public String toString() {
            return "DeviceEvent [phoneNumber=" + phoneNumber + ", bin=" + bin + ", bout=" + bout + ", timestamp="
                    + timestamp + "]";
        }
}
```

5. 设备架构序列化和反序列化

```
package com.book.flinkcep.example;

import java.io.IOException;

import org.apache.flink.api.common.typeinfo.TypeInformation;
import org.apache.flink.api.java.typeutils.TypeExtractor;
import org.apache.flink.streaming.util.serialization.DeserializationSchema;
import org.apache.flink.streaming.util.serialization.SerializationSchema;

public class DeviceSchema implements DeserializationSchema<DeviceEvent>,
SerializationSchema<DeviceEvent>{

    private static final long serialVersionUID = 1051444497161899607L;

    @Override
    public TypeInformation<DeviceEvent> getProducedType() {
        return TypeExtractor.getForClass(DeviceEvent.class);
    }
    @Override
    public byte[] serialize(DeviceEvent element) {
        return element.toString().getBytes();
    }
    @Override
    public DeviceEvent deserialize(byte[] message) throws IOException {
        return DeviceEvent.fromString(new String(message));
    }

    @Override
```

```java
        public boolean isEndOfStream(DeviceEvent nextElement) {
            return false;
        }
}
```

6. 以设备对象形式生成数据

最后,剩下的代码将数据变量转换为实数。

```java
 package com.book.flinkcep.example;

import java.util.Properties;
import java.util.concurrent.ThreadLocalRandom;

import org.apache.kafka.clients.producer.KafkaProducer;
import org.apache.kafka.clients.producer.ProducerRecord;

public class DataGenerator {
    public static void main(String args[]) {
        Properties properties = new Properties();
        properties.put("bootstrap.servers", "localhost:9092");
        properties.put("key.serializer",
"org.apache.kafka.common.serialization.StringSerializer");
        properties.put("value.serializer",
"org.apache.kafka.common.serialization.StringSerializer");
        properties.put("acks", "1");
        KafkaProducer<Integer, String> producer = new
KafkaProducer<Integer, String>(properties);
        int counter =0;
        int nbrOfEventsRequired = Integer.parseInt(args[0]);
        while (counter<nbrOfEventsRequired) {
            StringBuffer stream = new StringBuffer();
            long phoneNumber =
ThreadLocalRandom.current().nextLong(99999999501,
                                    99999999991);
            int bin = ThreadLocalRandom.current().nextInt(100000,9999999);
            int bout = ThreadLocalRandom.current().nextInt(100000,9999999);
            stream.append(phoneNumber);
            stream.append(",");
            stream.append(bin);
            stream.append(",");
            stream.append(bout);
            stream.append(",");
            stream.append(System.currentTimeMillis());

            System.out.println(stream.toString());
            ProducerRecord<Integer, String> data = new
```

```
            ProducerRecord<Integer, String>(
                                "device-data", stream.toString());
                    producer.send(data);
                    counter++;
            }
            producer.close();
    }
}
```

7. 在 Flink 上运行 Storm 拓扑

Storm 拓扑无须修改即可在 Flink 环境中运行。在进行更改时，只需记住以下几点：

- `StormSubmitter` 替换为 `FlinkSubmitter`；
- `NimbusClient` 和 `Client` 替换为 `FlinkClient`；
- `LocalCluster` 替换为 `FlinkLocalCluster`。

下面显示了该函数的代码。

8. Flink 和 Storm 集成的主类

```
package com.book.flink.diy;

import org.apache.flink.storm.api.FlinkTopology;

import backtype.storm.topology.TopologyBuilder;

public class FlinkStormExample {
    public static void main(String[] args) throws Exception {
          TopologyBuilder topologyBuilder = new TopologyBuilder();
          topologyBuilder.setSpout("spout", new FileSpout("/tmp/devicedata.txt"), 1);
          topologyBuilder.setBolt("parser", new ParserBolt(),
1).shuffleGrouping("spout");
          topologyBuilder.setBolt("tdrCassandra", new
TDRCassandraBolt("localhost", "tdr"), 1).shuffleGrouping("parser",
"tdrstream");
          FlinkTopology.createTopology(topologyBuilder).execute();
    }
}
```

9. Storm 中的文件 spout

```
package com.book.flink.diy;

import java.io.BufferedReader;
import java.io.FileReader;
```

```java
import java.io.IOException;
import java.util.Map;

import backtype.storm.spout.SpoutOutputCollector;
import backtype.storm.task.TopologyContext;
import backtype.storm.topology.OutputFieldsDeclarer;
import backtype.storm.topology.base.BaseRichSpout;
import backtype.storm.tuple.Fields;
import backtype.storm.tuple.Values;

public class FileSpout extends BaseRichSpout {
    private static final long serialVersionUID = -6167039596158642349L;
    private SpoutOutputCollector collector;
    private String fileName;
    private BufferedReader reader;

    public FileSpout(String fileName)
        this.fileName = fileName;
    }
    @Override
    public void open(Map conf, TopologyContext context, SpoutOutputCollector collector) {
        //fileName = (String) conf.get("file");
        this.collector = collector;

        try {
            reader = new BufferedReader(new FileReader(fileName));
        } catch (Exception e) {
            throw new RuntimeException(e);
        }
    }
    @Override
    public void nextTuple() {
      try {
            String line = reader.readLine();
            if (line != null) {
                collector.emit(new Values(line));
            }
        } catch (IOException e) {
            e.printStackTrace();
        }
    }

    @Override
    public void close() {
        try {
            reader.close();
        } catch (IOException e) {
```

```java
                e.printStackTrace();
            }
    }
    @Override
    public void declareOutputFields(OutputFieldsDeclarer declarer) {
        Fields schema = new Fields("line");
        declarer.declare(schema);
    }
}
```

10. Cassandra 持久化 bolt

```java
package com.book.flink.diy;

import java.util.Map;

import com.datastax.driver.core.Cluster;
import com.datastax.driver.core.Session;

import backtype.storm.task.TopologyContext;
import backtype.storm.topology.BasicOutputCollector;
import backtype.storm.topology.OutputFieldsDeclarer;
import backtype.storm.topology.base.BaseBasicBolt;
import backtype.storm.tuple.Tuple;

public class TDRCassandraBolt extends BaseBasicBolt {
    private static final long serialVersionUID = 1L;
    private Cluster cluster;
    private Session session;
    private String hostname;
    private String keyspace;

    public TDRCassandraBolt(String hostname, String keyspace) {
        this.hostname = hostname;
        this.keyspace = keyspace;
    }

    @Override
    public void prepare(Map stormConf, TopologyContext context) {
        cluster = Cluster.builder().addContactPoint(hostname).build();
        session = cluster.connect(keyspace);
    }
    public void execute(Tuple input, BasicOutputCollector arg1) {
        PacketDetailDTO packetDetailDTO = (PacketDetailDTO) input.getValueByField("tdrstream");
        System.out.println("field value "+ packetDetailDTO);
        session.execute("INSERT INTO packet_tdr (phone_number, bin, bout, timestamp) VALUES ("
```

```
                            + packetDetailDTO.getPhoneNumber()
                            + ", "
                            + packetDetailDTO.getBin()
                            + ","
                            + packetDetailDTO.getBout()
                            + "," + packetDetailDTO.getTimestamp() + ")");
    }
    public void declareOutputFields(OutputFieldsDeclarer arg0) {

    }

    @Override
    public void cleanup() {
        session.close();
        cluster.close();
    }
}
```

11. 在 bolt 中解析事件

```
package com.book.flink.diy;

import java.util.Map;

import backtype.storm.task.TopologyContext;
import backtype.storm.topology.BasicOutputCollector;
import backtype.storm.topology.OutputFieldsDeclarer;
import backtype.storm.topology.base.BaseBasicBolt;
import backtype.storm.tuple.Fields;
import backtype.storm.tuple.Tuple;
import backtype.storm.tuple.Values;

public class ParserBolt extends BaseBasicBolt {

    private static final long serialVersionUID = 1271439619204966337L;

    @Override
    public void prepare(Map stormConf, TopologyContext context) {
    }

    @Override
    public void execute(Tuple input, BasicOutputCollector collector) {
        String valueByField = input.getString(0);
        System.out.println("field value "+ valueByField);
        String[] split = valueByField.split(",");
        PacketDetailDTO tdrPacketDetailDTO = new PacketDetailDTO();
```

```java
        tdrPacketDetailDTO.setPhoneNumber(Long.parseLong(split[0]));
        tdrPacketDetailDTO.setBin(Integer.parseInt(split[1]));
        tdrPacketDetailDTO.setBout(Integer.parseInt(split[2]));
        tdrPacketDetailDTO.setTimestamp(Long.parseLong(split[3]));

        collector.emit("tdrstream", new Values(tdrPacketDetailDTO));
    }

    @Override
    public void cleanup() {
    }

    @Override
    public void declareOutputFields(OutputFieldsDeclarer declarer) {
        declarer.declareStream("tdrstream", new Fields("tdrstream"));
    }
}
```

12. bolt 之间的数据传输对象

```java
package com.book.flink.diy;

import java.io.Serializable;

public class PacketDetailDTO implements Serializable {

    private static final long serialVersionUID = 9148607866335518739L;
    private long phoneNumber;
    private int bin;
    private int bout;
    private int totalBytes;
    private long timestamp;

    public long getPhoneNumber() {
        return phoneNumber;
    }

    public void setPhoneNumber(long phoneNumber) {
        this.phoneNumber = phoneNumber;
    }

    public int getBin() {
        return bin;
    }
```

```java
    public void setBin(int bin) {
        this.bin = bin;
    }

    public int getBout() {
        return bout;
    }

    public void setBout(int bout) {
        this.bout = bout;
    }

    public int getTotalBytes() {
        return totalBytes;
    }

    public void setTotalBytes(int totalBytes) {
        this.totalBytes = totalBytes;
    }

    public long getTimestamp() {
        return timestamp;
    }

    public void setTimestamp(long timestamp) {
        this.timestamp = timestamp;
    }
}
```

12.10 小结

在本章，我们向读者介绍了 Flink 的体系结构，讨论了 Kappa 架构和 Flink 的工作原理。Flink 有不同的源和 sink。解释了 Kafka 和 RabbitMQ 等源的示例。以 Kafka 为源解释了 Cassandra 等作为 sink 的例子。Flink 分别为流处理和批处理提供了 DataSet 和 DataStream API。介绍了每个 API 可用的不同转换。Flink 提供了两个高级程序库：CEP 和 Gelly。CEP 通过模式实现进行实时处理，Gelly 是一个基于 Flink 的图 API。最后，留给了读者一些需要自己解决的问题。

在第 13 章，读者将了解如何使用本书所有场景的应用开发一个真正的示例。

第六部分 综合应用

- 第 13 章 用例研究

第 13 章
用例研究

在学完前面的章节之后,读者应该已经了解了可用于实时和批处理流的不同框架。在本章,我们研究的综合用例使用了前几章中讨论的框架案例。

---本章主要包括以下内容---

- 用例研究概述
- 数据建模
- 工具和框架
- 建立基础设施
- 实现用例
- 运行用例

13.1 概述

本节将讨论和实现 Geofencing 用例,并将其作为用例研究。Geofencing 通过**全球定位系统(GPS)的卫星网络**或本地射频标识符(如 Wi-Fi 节点或蓝牙信标)在一个位置周围创建虚拟边界。Geofencing 与软件/硬件配对,以检测虚拟边界并执行用户定义的适当操作。大多数 Geofencing 用例都是利用物联网来解决的。基于 Geofencing 的实时用例如下。

- 假设孩子开我们的车去市场购买商品,我们不希望他们离家超出 5 公里或 5 英里的距离,或者我们想在地图上了解孩子/其他家庭成员的位置以掌握他们的行踪。
- 当我们已将汽车去维修了,并希望汽车离汽车服务中心不超过 1 公里或 1 英里。

- 当我们接近家时，车库门会自动打开，不必下车，通过手动地推/拉打开它。
- 当我们离开家的时候，想着是不是把门锁上了。
- 假设有一个养牛场，希望牛的活动不要超出一定的距离。
- 基于信标的用例可以在商店或商场中实现，以了解客户并推荐他/她可能喜欢的东西。

除了前面提到的用例，还有更多的用例。在这里，我们选择了第一个提到的用例：车辆 Geofencing。如果车辆超出用户定义的虚拟边界，用户将收到警告。

在此用例中，我们将实时车辆定位数据推送到 Kafka 中，其将保存车辆的静态数据，包含车辆的起点和 Kafka 中的阈值距离，并把它们推送到 Hazelcast 中。实时车辆数据由 Storm 读取，然后它将开始处理每个车辆事件，检查当前位置和用户定义位置的距离，如果车辆位置超过阈值距离，则会生成警报。

13.2 数据建模

实时车辆数据传感器模型如表 13-1 所示。

表 13-1

字段名称	车辆 Id	纬度	经度	速度	时间戳
数据类型	String	Double	Double	Integer	Long

车辆 Id（即底盘号标识）是任何车辆的唯一标识符，通常，对于不同类型的车辆，底盘号标识在不同的位置。GPS 检测纬度和经度，告诉用户车辆的当前位置。速度是车辆的速度。时间戳是生成此事件的时间。

车辆静态数据模型如表 13-2 所示。

表 13-2

字段名称	车辆 Id	纬度	经度	距离	电话号码
数据类型	String	Double	Double	Double	String

此静态数据由车主在为车辆设置警报时提供。车辆 Id 是任何车辆的唯一标识符。纬

度和经度是车辆的起始位置或从计算距离到发出警报的位置。距离是以米为单位的阈值距离。电话号码在生成警报 Id 时发送通知。

输出被推送到 Elasticsearch 中。在这里将使用两种类型的数据模型,一种用于存储车辆传感器的实时数据,另一种用于存储系统生成的警报信息。

实时传感器数据模型如表 13-3 所示。

表 13-3

字段名称	坐标	速度	时间戳	车辆 Id
数据类型	Geo_point	Integer	Date	Text

坐标包含 JSON 格式的纬度和经度,该格式由 Elasticsearch 转换为 geo_point 类型。速度是车辆的速度。时间戳是事件发生的时间。车辆 Id 是车辆的唯一标识符。

警报信息数据模型如表 13-4 所示。

表 13-4

字段名称	实际坐标	期望距离	实际距离	期望坐标	时间戳	车辆 Id
数据类型	Geo_point	Double	Double	Geo_point	Date	Text

实际坐标包含车辆实时的当前纬度和经度并作为位置。期望距离是用户为车辆配置的距离阈值。实际距离是实际坐标和预期坐标之间的当前距离。期望坐标是用户在设置车辆警报时配置的起点或位置。时间戳是系统生成警报的时间。车辆 Id 是车辆的唯一标识符。

Hazelcast 将创建两种类型的 Map 以实时处理事件:VehicleAlertInfo 和 GeneratedAlerts。在 VehicleAlertInfo 中,Map 键是车辆 Id,值包含车辆 Id、纬度、经度、距离和电话号码的 Java 对象。GeneratedAlerts 的映射键是车辆 Id,值包含阈值距离、实际距离、起始纬度、起始经度、实际纬度、实际经度、车辆 Id、时间戳和电话号码。

13.3 工具和框架

表 13-5 所示的工具和框架用于实现完整的用例。

表 13-5

名称	版本
Java	1.8
Zookeeper	3.4.6
Kafka	2.11-0.8.2.2
Hazelcast	3.8
Storm	1.1.1
Elasticsearch	5.2.2
Kibana	5.2.2

13.4 建立基础设施

要实现此用例，必须设置以下工具。

- **Hazelcast**：在前面的章节中，我们讨论了使用 Java 代码并在 Eclipse 中运行 Hazelcast 的设置。现在将讨论使用脚本运行 Hazelcast。首先下载 Hazelcast 的设置。解压后，用户将获得图 13-1 所示的文件夹和文件。

图 13-1

在 `hazelcast.xml` 中进行更改以启用 Hazelcast UI 作为 `mancenter`。

```
<management-center enabled="true">
http://localhost:8080/mancenter</management-center >
```

现在，执行以下脚本来启动 Hazelcast。

```
/bin/start.sh
```

这将在本地主机上启动 Hazelcast 并绑定到端口 5701。如果想为 Hazelcast 创建一个集群，那么把 Hazelcast 安装目录复制到不同的位置，然后再次执行 `start.sh` 脚本。它会在本地主机上以集群模式启动 Hazelcast，并将第二个实例与端口 5702 绑定。

要启动 mancenter UI，请执行以下脚本：

```
/mancenter/startMancenter.sh
```

它将在本地主机的端口 8080 上启动 mancenter UI。URL 是 http://localhost:8080/mancenter/。它会在 mancenter 目录中自动创建一个工作目录。如果要在其他端口和其他位置的工作目录上启动 UI，请执行以下命令：

```
/mancenter/startMancenter.sh <PORT> <PATH>
```

mancenter UI 如图 13-2 所示。

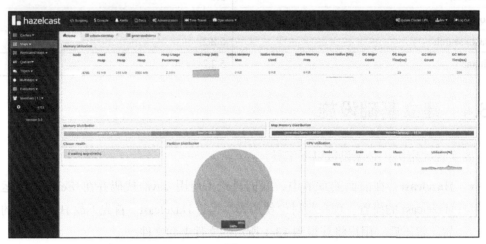

图 13-2

（1）**使用 WhatsApp 通知**：为了在 WhatsApp 上发送通知，需要一个可以发送和接收消息的库，比如 Yowsup。在系统上设置 Yowsup 的步骤如下。

① 使用以下命令克隆 git 仓库 `Yowsup`。

```
git clone git://github.com/tgalal/yowsup.git
```

② 在 `yowsup/env/env_android.py` 中更改以下属性。

```
_MD5_CLASSES = "ox998VW0nBTueMVfjuZkmQ=="
_VERSION = "2.17.212"
```

③ **使用 WhatsApp 注册用户的号码**：使用未曾在 WhatsApp 上使用过的号码进行注册，因为不能再次注册相同的号码。使用以下命令进行注册：

```
yowsup-cli registration -d -r sms -C <Country Code> -p
<Phone number with country code and without '+'> -m <MCC> -
n <MNC> -E android
```

执行上述步骤后，用户的号码将获得 6 位验证码，并在控制台上得到图 13-3 所示的响应。

```
INFO:yowsup.common.http.warequest:{"login":"         ","status":"sent","length":6,"method":"sms","retry_after":65,"sms_wait":65,"voice_wait":5}
status: sent
retry_after: 65
length: 6
login:
method: sms
```

图 13-3

使用以下命令通过验证码注册用户的号码。

```
yowsup-cli registration -d -R 705-933 -p 917988141683 -C 91
-E android
```

用户将在控制台上获得图 13-4 所示的响应。

图 13-4

④ 现在，创建一个具有以下属性的配置文件。

```
cc=<Country Code>
phone=<Phone number with country code and without '+'>
password=<Password that you received in previous step>
```

将其另存为 **whatsapp_config.txt**。（可以给它取任何你喜欢的名字，并在下一步中相应地更新文件名。）

⑤ 注册后，可以使用以下命令发送消息进行测试：

```
yowsup-cli demos --config whatsapp_config.txt --send
<Receiver Phone number with country code and without '+'>
"<Your message>"
```

检查手机，应该已经收到消息。

（2）Zookeeper：Kafka 和 Storm 需要 Zookeeper。Zookeeper 可以从 Kafka 设置中启动。已经在第 3 章详细讨论了 Kafka 的设置。所以，现在可以使用以下命令启动它：

```
/bin/zookeeper-server-start.sh ../config/zookeeper.properties
```

（3）**Kafka**：使用以下命令启动 Kafka。

```
/bin/kafka-server-start.sh ../config/server.properties
```

使用以下命令在 Kafka 上创建两个主题：`vehicle-data` 和 `vehicle-static-data`。

```
/bin/kafka-topics.sh --create --topic vehicle--data --zookeeper localhost:2181 --partitions 1 --replication-factor 1
/bin/kafka-topics.sh --create --topic vehicle-static-data --zookeeper localhost:2181 --partitions 1 --replication-factor 1
```

（4）**Elasticsearch**：已经在第 3 章详细讨论了 Elasticsearch 的设置。使用以下命令启动它：

```
/bin/elasticsearch
```

在 Elasticsearch 中创建两个索引：`tdr` 和 `alert`。以下命令用于创建它们。

```
curl -XPUT 'localhost:9200/vehicle-tdr?pretty&pretty'
curl -XPUT 'localhost:9200/vehicle-tdr/_mapping/tdr' -d '{
        "properties": {
            "coords": {
                "type": "geo_point"
            },
            "speed": {
                "type": "integer"
            },
            "timestamp": {
                "type": "date"
            },
            "vehicle_id" : {
                "type" : "text",
                "fields" : {
                    "keyword" : {
                        "type" : "keyword",
                        "ignore_above" : 256
                    }
                }
            }
        }
    }'
curl -XPUT 'localhost:9200/vehicle-alert?pretty&pretty'
curl -XPUT 'localhost:9200/vehicle-alert/_mapping/alert' -d
'{
        "properties": {
            "actual_coords": {
```

```
        "type": "geo_point"
      },
      "expected_distance": {
        "type": "double"
      },
      "actual_distance": {
        "type": "double"
      },
      "expected_coords": {
        "type": "geo_point"
      },
      "timestamp": {
        "type": "date"
      },
      "vehicle_id" : {
          "type" : "text",
          "fields" : {
            "keyword" : {
              "type" : "keyword",
              "ignore_above" : 256
            }
          }
       }
    }
}'
```

（5）**Kibana**：下载 Kibana。这里将下载 Kibana 的 5.2.2 版本，与 Elasticsearch 相同。解压后，将获得图 13-5 所示的文件夹和文件。

```
:~/demo/kibana-5.2.2-linux-x86_64$ ls
bin  config  data  LICENSE.txt  node  node_modules  NOTICE.txt  optimize  package.json  plugins  README.txt  src  ui_framework  webpackshims
```

图 13-5

要启动 Kibana，请执行以下命令：

```
/bin/ kibana
```

Kibana UI 可以使用 URL `http://localhost:5601` 来访问。

（6）**Storm**：在第 4 章中已经详细讨论了 Storm 的设置和配置。因此，现在启动 `nimbus`、`supervisor`、`ui` 和 `logviewer` 服务。由于 Hazelcast mancenter UI 在 8080 上运行，因此需要将 Storm UI 的端口更改为在端口 8081 上运行，在 `storm.yaml` 中添加以下行：

```
ui.port: 8081
```

通过执行以下命令启动 Storm 服务。

（7）**Nimbus**：：首先，需要在 Storm 中启动 Nimbus 服务。执行以下命令启动它。

```
/bin/storm nimbus
```

（8）**Supervisor**：接下来，需要启动 Supervisor 节点来连接 Nimbus 节点。执行以下命令启动它。

```
/bin/storm supervisor
```

（9）**UI**：执行以下命令启动 Storm UI。

```
/bin/storm ui
```

可以在 `http://nimbus-host:8081` 中访问 UI。

（10）**Logviewer**：日志查看器服务帮助查看 Storm UI 上的工作单元日志。执行以下命令启动它。

```
/bin/storm logviewer
```

13.5 实现用例

为了实现 geofencing 用例，我们可以用以下组件进行构建和开发。

13.5.1 构建数据模拟器

在生成车辆的实时数据之前，需要有车辆的起始点。以下是生成这些实时数据的代码：

```
package com.book.simulator;
import java.util.HashMap;
import java.util.Map;
import java.util.Properties;
import java.util.Random;
import org.apache.kafka.clients.producer.KafkaProducer;
import org.apache.kafka.clients.producer.ProducerRecord;
import org.apache.kafka.common.serialization.StringSerializer;
import com.book.domain.Location;
import com.fasterxml.jackson.core.JsonProcessingException;
import com.fasterxml.jackson.databind.ObjectMapper;
/**
* This class is used to generate vehicle start point for number of vehicle
* specified by user.
```

```java
 *
 * @author SGupta
 *
 */
public class VehicleStartPointGenerator {
    static private ObjectMapper objectMapper = new ObjectMapper();
    static private Random r = new Random();
    static private String BROKER_1_CONNECTION_STRING = "localhost:9092";
    static private String KAFKA_TOPIC_STATIC_DATA = "vehicle-static-data";
    public static void main(String[] args) {
      if (args.length < 1) {
        System.out.println("Provide number of vehicle");
        System.exit(1);
      }
      // Number of vehicles for which data needs to be generated.
      int numberOfvehicle = Integer.parseInt(args[0]);
      // Get producer to push data into Kafka
      KafkaProducer<Integer, String> producer = configureKafka();
      // Get vehicle start point.
      Map<String, Location> vehicleStartPoint =
getVehicleStartPoints(numberOfvehicle);
 // Push data into Kafka
pushVehicleStartPointToKafka(vehicleStartPoint, producer);
producer.close();
}
    private static KafkaProducer<Integer, String> configureKafka() {
      Properties properties = new Properties();
      properties.put("bootstrap.servers", BROKER_1_CONNECTION_STRING);
      properties.put("key.serializer", StringSerializer.class.getName());
      properties.put("value.serializer",StringSerializer.class.getName());
      properties.put("acks", "1");
      KafkaProducer<Integer, String> producer = new KafkaProducer<Integer,
String>(properties);
      return producer;
    }
    private static Map<String, Location> getVehicleStartPoints(int
numberOfvehicle) {
      Map<String, Location> vehicleStartPoint = new HashMap<String,
Location>();
      for (int i = 1; i <= numberOfvehicle; i++) {
        vehicleStartPoint.put("v" + i,
        new Location((r.nextDouble() * -180.0) + 90.0, (r.nextDouble() *
-360.0) + 180.0));
      }
      System.out.println(vehicleStartPoint);
```

```
      return vehicleStartPoint;
  }
  private static void pushVehicleStartPointToKafka(Map<String, Location>
vehicleStartPoint,
    KafkaProducer<Integer, String> producer) {
      ProducerRecord<Integer, String> data = null;
      try {
        data = new ProducerRecord<Integer, String>(KAFKA_TOPIC_STATIC_DATA,
          objectMapper.writeValueAsString(vehicleStartPoint));
        } catch (JsonProcessingException e) {
        e.printStackTrace();
      }
      producer.send(data);
    }
  }
```

现在需要构建一个实时数据模拟器,在用户指定值的半径内生成车辆数据:

```
package com.book.simulator;

import java.io.IOException;
import java.util.HashMap;
import java.util.Map;
import java.util.Properties;
import java.util.Random;

import org.apache.kafka.clients.producer.KafkaProducer;
import org.apache.kafka.clients.producer.ProducerRecord;
import org.apache.kafka.common.serialization.StringDeserializer;
import org.apache.kafka.common.serialization.StringSerializer;

import com.book.domain.Location;
import com.book.domain.VehicleSensor;
import com.fasterxml.jackson.core.JsonProcessingException;
import com.fasterxml.jackson.core.type.TypeReference;
import com.fasterxml.jackson.databind.ObjectMapper;

import kafka.consumer.Consumer;
import kafka.consumer.ConsumerConfig;
import kafka.consumer.ConsumerIterator;
import kafka.consumer.KafkaStream;
import kafka.javaapi.consumer.ConsumerConnector;
/**
 * This class is used to generate real-time vehicle data with updated
location
```

```
 * within distance in radius of user specified value. Messages are pushed
 into Kafka topic.
 *
 * @author SGupta
 *
 */
public class VehicleDataGeneration {

    static private ObjectMapper objectMapper = new ObjectMapper();
    static private Random r = new Random();
    static private String BROKER_1_CONNECTION_STRING = "localhost:9092";
    static private String ZOOKEEPER_CONNECTION_STRING = "localhost:2181";
    static private String KAFKA_TOPIC_STATIC_DATA = "vehicle-static-data";
    static private String KAFKA_TOPIC_REAL_TIME_DATA = "vehicle-data";

    public static void main(String[] args) {
        if (args.length < 2) {
            System.out.println("Provide total number of records and
range of distance from start point");
            System.exit(1);
        }

        //Total number of records this simulator will generate
        int totalNumberOfRecords = Integer.parseInt(args[0]);
        //Distance in meters as Radius
        int distanceFromVehicleStartPoint = Integer.parseInt(args[1]);

        // Get Kafka producer
        KafkaProducer<Integer, String> producer = configureKafka();

        // Get Vehicle Start Points
        Map<String, Location> vehicleStartPoint = getVehicleStartPoints();
        // Generate data within distance and push to Kafka
        generateDataAndPushToKafka(producer, vehicleStartPoint.size(),
totalNumberOfRecords,
                    distanceFromVehicleStartPoint, vehicleStartPoint);
        producer.close();
    }

    private static KafkaProducer<Integer, String> configureKafka() {
        Properties properties = new Properties();

        properties.put("bootstrap.servers", BROKER_1_CONNECTION_STRING);
        properties.put("key.serializer",
StringSerializer.class.getName());
        properties.put("value.serializer",
```

```java
        StringSerializer.class.getName());
        properties.put("acks", "1");

        KafkaProducer<Integer, String> producer = new
KafkaProducer<Integer, String>(properties);
        return producer;
    }

    private static void generateDataAndPushToKafka(KafkaProducer<Integer,
String> producer, int numberOfvehicle,
            int totalNumberOfRecords, int distanceFromVehicleStartPoint,
Map<String, Location> vehicleStartPoint) {
        for (int i = 0; i < totalNumberOfRecords; i++) {
            int vehicleNumber = r.nextInt(numberOfvehicle);
            String vehicleId = "v" + (vehicleNumber + 1);
            Location currentLocation = vehicleStartPoint.get(vehicleId);
            Location locationInLatLngRad =
getLocationInLatLngRad(distanceFromVehicleStartPoint, currentLocation);
            System.out.println(
                    "Vehicle number is " + vehicleId + " and
location is " + locationInLatLngRad + " with distance of "
                            +
(getDistanceFromLatLonInKm(currentLocation.getLatitude(),
currentLocation.getLongitude(),
locationInLatLngRad.getLatitude(), locationInLatLngRad.getLongitude()) *
1000));

            VehicleSensor vehicleSensor = new VehicleSensor(vehicleId,
locationInLatLngRad.getLatitude(),
                    locationInLatLngRad.getLongitude(),
r.nextInt(100), System.currentTimeMillis());

            ProducerRecord<Integer, String> data = null;
            try {
                data = new ProducerRecord<Integer,
String>(KAFKA_TOPIC_REAL_TIME_DATA,
objectMapper.writeValueAsString(vehicleSensor));
            } catch (JsonProcessingException e) {
                e.printStackTrace();
            }

            try {
                Thread.sleep(1000);
            } catch (InterruptedException e) {
                e.printStackTrace();
```

```
            }
                producer.send(data);
        }
    }

    private static Map<String, Location> getVehicleStartPoints() {
        Map<String, Location> vehicleStartPoint = new HashMap<String, Location>();
        Properties props = new Properties();
        props.put("zookeeper.connect", ZOOKEEPER_CONNECTION_STRING);
        props.put("group.id", "DataLoader" + r.nextInt(100));
        props.put("key.deserializer", StringDeserializer.class.getName());
        props.put("value.deserializer", StringDeserializer.class.getName());

        ConsumerConnector consumer =
Consumer.createJavaConsumerConnector(new ConsumerConfig(props));

        Map<String, Integer> topicCountMap = new HashMap<String, Integer>();
        topicCountMap.put(KAFKA_TOPIC_STATIC_DATA, new Integer(1));

        KafkaStream<byte[], byte[]> stream =
consumer.createMessageStreams(topicCountMap).get(KAFKA_TOPIC_STATIC_DATA)
                .get(0);

        ConsumerIterator<byte[], byte[]> it = stream.iterator();

        while (it.hasNext()) {
            String message = new String(it.next().message());
            try {
                vehicleStartPoint = objectMapper.readValue(message,
new TypeReference<Map<String, Location>>() {
                });
            } catch (IOException e) {
                e.printStackTrace();
            }
            break;
        }
        consumer.shutdown();
        return vehicleStartPoint;
    }

    public static double getDistanceFromLatLonInKm(double lat1, double lon1,double la
```

第13章 用例研究

```java
t2, double lon2) {
        int R = 6371; // Radius of the earth in km
        double dLat = deg2rad(lat2 - lat1); // deg2rad below
        double dLon = deg2rad(lon2 - lon1);
        double a = Math.sin(dLat / 2) * Math.sin(dLat / 2)
                 + Math.cos(deg2rad(lat1)) * Math.cos(deg2rad(lat2)) *
Math.sin(dLon / 2) * Math.sin(dLon / 2);
        double c = 2 * Math.atan2(Math.sqrt(a), Math.sqrt(1 - a));
        double d = R * c; // Distance in km
        return d;
    }

    private static double deg2rad(double deg) {
        return deg * (Math.PI / 180);
    }

    protected static Location getLocationInLatLngRad(double radiusInMeters,
Location currentLocation) {
        double x0 = currentLocation.getLongitude();
        double y0 = currentLocation.getLatitude();

        Random random = new Random();

        // Convert radius from meters to degrees.
        double radiusInDegrees = radiusInMeters / 111320f;

        // Get a random distance and a random angle.
        double u = random.nextDouble();
        double v = random.nextDouble();
        double w = radiusInDegrees * Math.sqrt(u);
        double t = 2 * Math.PI * v;
        // Get the x and y delta values.
        double x = w * Math.cos(t);
        double y = w * Math.sin(t);

        // Compensate the x value.
        double new_x = x / Math.cos(Math.toRadians(y0));

        double foundLatitude;
        double foundLongitude;

        foundLatitude = y0 + y;
        foundLongitude = x0 + new_x;

        Location copy = new Location(currentLocation);
```

```
            copy.setLatitude(foundLatitude);
            copy.setLongitude(foundLongitude);
            return copy;
    }
}
```

13.5.2　Hazelcast 加载器

现在需要将车辆的静态值连同用户的电话号码添加到 Hazelcast 中，以便在处理事件时，Storm 能够计算车辆起点和当前位置之间的距离。代码如下：

```
package com.book.simulator;
import java.io.IOException;
import java.util.HashMap;
import java.util.Map;
import java.util.Properties;
import java.util.Random;
import org.apache.kafka.common.serialization.StringDeserializer;
import com.book.domain.Location;
import com.book.domain.VehicleAlertInfo;
import com.fasterxml.jackson.core.JsonParseException;
import com.fasterxml.jackson.core.type.TypeReference;
import com.fasterxml.jackson.databind.JsonMappingException;
import com.fasterxml.jackson.databind.ObjectMapper;
import com.hazelcast.client.HazelcastClient;
import com.hazelcast.client.config.ClientConfig;
import com.hazelcast.core.HazelcastInstance;
import kafka.consumer.Consumer;
import kafka.consumer.ConsumerConfig;
import kafka.consumer.ConsumerIterator;
import kafka.consumer.KafkaStream;
import kafka.javaapi.consumer.ConsumerConnector;
/**
* This class is used to load static information in to Hazelcast like 1. Map
1
* contains information of vehicle and its owner. 2. Map 2 contains
information
* of geo fencing range configured for each vehicle by user
*
* @author Sgupta
*
*/
    public class HazelCastLoader {
        static private Random r = new Random();
        static private ObjectMapper objectMapper = new ObjectMapper();
```

第 13 章　用例研究

```java
    public static void main(String[] args) {
      if (args.length < 3) {
        System.out.println("Provide phonenumber, topic and threshold distance for each vehicle");
        System.exit(1);
      }
      // Phone number on which user will get alert.
      String phoneNUmber = args[0];
      // Topic Name from which static data will be read and feed into Hazelcast
      String topic = args[1];
      // Threshold distance in meters.
      int distanceFromVehicleStartPoint = Integer.parseInt(args[2]);
      // Get vehicle alert info map from Hazelcast
      Map<String, VehicleAlertInfo> vehicleAlertMap = getHCAlertInfoMap();
      // Get message from Kafka and push into Hazelcast
      getAndLoadHCMap(phoneNUmber, topic, distanceFromVehicleStartPoint, vehicleAlertMap);
    }
  private static void getAndLoadHCMap(String phoneNUmber, String topic, int distanceFromVehicleStartPoint,
    Map<String, VehicleAlertInfo> vehicleAlertMap) {
      Properties props = new Properties();
      props.put("zookeeper.connect", "localhost:2181");
      props.put("group.id", "myGroup"+r.nextInt(100));
      props.put("key.deserializer", StringDeserializer.class.getName());
      props.put("value.deserializer", StringDeserializer.class.getName());
ConsumerConnector consumer = Consumer.createJavaConsumerConnector(new ConsumerConfig(props));
      Map<String, Integer> topicCountMap = new HashMap<String, Integer>();
      topicCountMap.put(topic, new Integer(1)); // number of consumer threads
      KafkaStream<byte[], byte[]> stream =
consumer.createMessageStreams(topicCountMap).get(topic).get(0);
      ConsumerIterator<byte[], byte[]> it = stream.iterator();
      while(it.hasNext()) {
        try {
          String message = new String(it.next().message());
          System.out.println("Message: "+ message);
          Map<String, Location> readValue = objectMapper.readValue(message,
          new TypeReference<Map<String, Location>>() {
          });
          for (String vehicleId : readValue.keySet()) {
          VehicleAlertInfo vehicleAlertInfo = new VehicleAlertInfo();
          vehicleAlertInfo.setVehicleId(vehicleId);
vehicleAlertInfo.setLatitude(readValue.get(vehicleId).getLatitude());
```

```
vehicleAlertInfo.setLongitude(readValue.get(vehicleId).getLongitude());
        vehicleAlertInfo.setDistance(distanceFromVehicleStartPoint);
         vehicleAlertInfo.setPhoneNumber(phoneNUmber);
        vehicleAlertMap.put(vehicleId, vehicleAlertInfo);}
    } catch (JsonParseException e) {
        e.printStackTrace();
    } catch (JsonMappingException e) {
        e.printStackTrace();
    } catch (IOException e) {
        e.printStackTrace();\
    } catch (Exception e1) {
     e1.printStackTrace();
     }
    }
  consumer.shutdown();
  }
  private static Map<String, VehicleAlertInfo> getHCAlertInfoMap() {
    HazelcastInstance client = getHazelcastClient();
    Map<String, VehicleAlertInfo> vehicleAlertMap =
client.getMap("vehicleAlertMap");
    vehicleAlertMap.clear();
    return vehicleAlertMap;
  }
  private static HazelcastInstance getHazelcastClient() {
    ClientConfig clientConfig = new ClientConfig();
    return HazelcastClient.newHazelcastClient(clientConfig);
  }
}
```

13.5.3 构建 Storm 拓扑

构建一个完整的拓扑结构需要不同的工作单元。还需要一个 spout 作为拓扑的起点，在本例中为 `KafkaSpout`。可使用 Storm 提供的 Storm-Kafka 预构建 API。spout 之后就是此用例所需的组件。

1. 解析 bolt

这个 bolt 用于读取来自 Kafka spout 的消息，并将其囊括到 Java POJO 类中。代码如下：

```
package com.book.processing;
import java.io.IOException;
import java.util.Map;
```

```java
import org.apache.storm.task.TopologyContext;
import org.apache.storm.topology.BasicOutputCollector;
import org.apache.storm.topology.OutputFieldsDeclarer;
import org.apache.storm.topology.base.BaseBasicBolt;
import org.apache.storm.tuple.Fields;
import org.apache.storm.tuple.Tuple;
import org.apache.storm.tuple.Values;
import com.book.domain.VehicleSensor;
import com.fasterxml.jackson.core.JsonParseException;
import com.fasterxml.jackson.databind.JsonMappingException;
import com.fasterxml.jackson.databind.ObjectMapper;
/**
* This bolt is used to parse Vehicle real-time data from Kafka and convert it
* into {@link VehicleSensor} POJO object
*
* @author SGupta
*
*/
public class ParseBolt extends BaseBasicBolt {
  private static final long serialVersionUID = -2557041273635037199L;
  ObjectMapper objectMapper;
  @Override
  public void prepare(Map stormConf, TopologyContext context) {
    objectMapper = new ObjectMapper();
  }
  @Override
  public void execute(Tuple input, BasicOutputCollector collector) {
    // Take default message from KafkaSpout
    String valueByField = input.getString(0);
    VehicleSensor vehicleSensor = null;
    try {
      //Covert JSON value into VehicleSensor object
      vehicleSensor = objectMapper.readValue(valueByField,
VehicleSensor.class);
    } catch (JsonParseException e) {
      e.printStackTrace();
    } catch (JsonMappingException e) {
      e.printStackTrace();
    } catch (IOException e) {
      e.printStackTrace();
    }
    // Emit tuple to next bolt with steamId as parsedstream\
    collector.emit("parsedstream", new Values(vehicleSensor));
  }
```

13.5 实现用例

```
@Override
public void declareOutputFields(OutputFieldsDeclarer declarer) {
   // Declare Stream with streamId as parsedstream with fields parsedstream
   declarer.declareStream("parsedstream", new Fields("parsedstream"));
  }
}
```

2. 检查距离和警报 bolt

这个 bolt 检查从 Hazelcast 中读取的车辆起始位置与解析 bolt 后接收到的当前位置之间的距离。如果当前距离大于阈值距离，则将元组发送到下一个 bolt 以处理警报。代码如下：

```java
package com.book.processing;
import java.util.Map;
import org.apache.storm.task.TopologyContext;
import org.apache.storm.topology.BasicOutputCollector;
import org.apache.storm.topology.OutputFieldsDeclarer;
import org.apache.storm.topology.base.BaseBasicBolt;
import org.apache.storm.tuple.Fields;
import org.apache.storm.tuple.Tuple;
import org.apache.storm.tuple.Values;
import com.book.domain.AlertEvent;
import com.book.domain.VehicleAlertInfo;
import com.book.domain.VehicleSensor;
import com.hazelcast.client.HazelcastClient;
import com.hazelcast.client.config.ClientConfig;
import com.hazelcast.core.HazelcastInstance;
/**
* This bolt is used to first check distance between start point of vehicle and
* current location of vehicle. Only those tuples are emitted which are related
* to alerts.
*
* @author SGupta
*
*/
public class CheckDistanceAndAlertBolt extends BaseBasicBolt {
  private static final long serialVersionUID = -8873075873347212209L;
  private Map<String, VehicleAlertInfo> vehicleAlertMap;
  private HazelcastInstance hazelcastClient;
  private String host;
```

```java
    private String port;
    public CheckDistanceAndAlertBolt(String host, String port) {
      this.host = host;
      this.port = port;
    }
    @Override
    public void prepare(Map stormConf, TopologyContext context) {
      ClientConfig clientConfig = new ClientConfig();
      clientConfig.getNetworkConfig().addAddress(host + ":" + port);
      hazelcastClient = HazelcastClient.newHazelcastClient(clientConfig);
      vehicleAlertMap = hazelcastClient.getMap("vehicleAlertMap");
    }
    @Override
    public void execute(Tuple input, BasicOutputCollector collector) {
      // Read input with field name as parsedstream
      VehicleSensor vehicleSensor = (VehicleSensor)
input.getValueByField("parsedstream");
      String vehicleId = vehicleSensor.getVehicleId();
      // Get vehicle alert information from Hazelcast
      VehicleAlertInfo vehicleAlertInfo = vehicleAlertMap.get(vehicleId);
      // Get the distance between starting location and current location.
      double actualDistance =
getDistanceFromLatLonInKm(vehicleAlertInfo.getLatitude(),
        vehicleAlertInfo.getLongitude(), vehicleSensor.getLatitude(),
vehicleSensor.getLongitude()) * 1000;
      long thresholdDistance = vehicleAlertInfo.getDistance();
      // If current distance is more than threshold distance then emit tuple
      // to next bolt
      if (actualDistance > thresholdDistance) {
        AlertEvent alertEvent = new AlertEvent();
        alertEvent.setActualDistance(actualDistance);
        alertEvent.setThresholdDistance(thresholdDistance);
        alertEvent.setStartingLatitude(vehicleAlertInfo.getLatitude());
        alertEvent.setStartingLongitude(vehicleAlertInfo.getLongitude());
        alertEvent.setActualLatitude(vehicleSensor.getLatitude());
        alertEvent.setActualLongitude(vehicleSensor.getLongitude());
        alertEvent.setVehicleId(vehicleId);
        alertEvent.setTimeStamp(System.currentTimeMillis());
      alertEvent.setPhoneNumber(vehicleAlertInfo.getPhoneNumber());
      collector.emit("alertInfo", new Values(alertEvent));}
    }
    @Override
    public void declareOutputFields(OutputFieldsDeclarer declarer) {
      declarer.declareStream("alertInfo", new Fields("alertInfo"));
    }
```

```
@Override
public void cleanup() {
    hazelcastClient.shutdown();
}
public static double getDistanceFromLatLonInKm(double lat1, double lon1,
double lat2, double lon2) {
    int R = 6371; // Radius of the earth in km
    double dLat = deg2rad(lat2 - lat1); // deg2rad below
    double dLon = deg2rad(lon2 - lon1);
    double a = Math.sin(dLat / 2) * Math.sin(dLat / 2)
    + Math.cos(deg2rad(lat1)) * Math.cos(deg2rad(lat2)) * Math.sin(dLon /
2) * Math.sin(dLon / 2);
    double c = 2 * Math.atan2(Math.sqrt(a), Math.sqrt(1 - a));
    double d = R * c; // Distance in km
    return d;
}
private static double deg2rad(double deg) {
    return deg * (Math.PI / 180);
}
}
```

3. 生成警报 bolt

这个 bolt 使用 **Twilio** 应用程序生成 SMS 形式的警报。此外，它会发射元组以便将其保存到 Elasticsearch 中。代码如下：

```
package com.book.processing;
import java.io.BufferedReader;
import java.io.InputStreamReader;
import java.math.BigDecimal;
import java.math.RoundingMode;
import java.util.Map;
import org.apache.storm.task.TopologyContext;
import org.apache.storm.topology.BasicOutputCollector;
import org.apache.storm.topology.OutputFieldsDeclarer;
import org.apache.storm.topology.base.BaseBasicBolt;
import org.apache.storm.tuple.Fields;
import org.apache.storm.tuple.Tuple;
import org.apache.storm.tuple.Values;
import com.book.domain.AlertEvent;
import com.hazelcast.client.HazelcastClient;
import com.hazelcast.client.config.ClientConfig;
import com.hazelcast.core.HazelcastInstance;
/**
 * This bolt is used to generate alert in form of SMS using Twilio
```

```
  application.
* Also emit tuples so that it can be saved into Elasticsearch.
*
* @author SGupta
*
*/
public class GenerateAlertBolt extends BaseBasicBolt {
  private static final long serialVersionUID = -6802250427993673417L;
  private Map<String, AlertEvent> vehicleAlertMap;
  private HazelcastInstance hazelcastClient;
  private String host;
  private String port;
  public GenerateAlertBolt(String host, String port) {
    this.host = host;
    this.port = port;
  }
  @Override
  public void prepare(@SuppressWarnings("rawtypes") Map stormConf,
TopologyContext context) {
    ClientConfig clientConfig = new ClientConfig();
    clientConfig.getNetworkConfig().addAddress(host + ":" + port);
    hazelcastClient = HazelcastClient.newHazelcastClient(clientConfig);
    vehicleAlertMap = hazelcastClient.getMap("generatedAlerts");
  }
  public void execute(Tuple input, BasicOutputCollector collector) {
    // Get alert event from checkDistanceAndAlert bolt.
    AlertEvent alertEvent = (AlertEvent)
input.getValueByField("alertInfo");
    // Reading map containing alerts from Hazelcast
    AlertEvent previousAlertEvent =
vehicleAlertMap.get(alertEvent.getVehicleId());
    // Check whether alert is already generated for this vehicle or not.
    if (previousAlertEvent == null) {
      // Add entry in Hazelcast Map
      vehicleAlertMap.put(alertEvent.getVehicleId(), alertEvent);
      System.out.println(alertEvent.toString());
      String message = "ALERT!! Hi, your vehicle id " +
alertEvent.getVehicleId()
          + " is moving out of start location i.e. ("
          + BigDecimal.valueOf(alertEvent.getActualLatitude()).setScale(2,
RoundingMode.HALF_DOWN)
            .doubleValue()
          + ","
          + BigDecimal.valueOf(alertEvent.getActualLongitude()).setScale(2,
RoundingMode.HALF_DOWN
```

```
      .doubleValue()
      + ") with distance " +
BigDecimal.valueOf(alertEvent.getActualDistance())
      .setScale(2, RoundingMode.HALF_DOWN).doubleValue();
    System.out.println(" GenerateAlertBOLT: >>>>> " + message);
    // Generate SMS.
    sendMessage(alertEvent.getPhoneNumber(), message);
    // Emit tuple for next bolt.
    collector.emit("generatedAlertInfo", new Values(alertEvent));
  } else {
    System.out.println(" GenerateAlertBOLT: >>>>> Alert is already 
generated for " +
    alertEvent.getVehicleId());
  }
 }
 public void declareOutputFields(OutputFieldsDeclarer declarer) {
   declarer.declareStream("generatedAlertInfo", new 
Fields("generatedAlertInfo"));
 }
 public void sendMessage(String phoneNumber, String message) {
   String s;
   Process p;
   try {
     String[] sendCommand = { "python", "yowsup-cli", "demos", "-c",
     "whatsapp_config.txt", "-s", phoneNumber, message };
     p = Runtime.getRuntime().exec(sendCommand);
     BufferedReader br = new BufferedReader(new 
InputStreamReader(p.getInputStream()));
     while ((s = br.readLine()) != null)
     System.out.println("line: " + s);
     BufferedReader errBr = new BufferedReader(new 
InputStreamReader(p.getErrorStream()));
     while ((s = errBr.readLine()) != null)
     System.out.println("line: " + s);
     p.waitFor();
     System.out.println("exit: " + p.exitValue());
     p.destroy();
   } catch (Exception e) {
   }
 }
}
```

4. Elasticsearch Bolt

这个 bolt 是一个持久性的 bolt。实时传感器数据或生成的警报将持久化到 Elasticsearch 中,以便在 Kibana 中进行进一步分析。代码如下:

第 13 章 用例研究

```java
package com.book.processing;
import java.net.InetSocketAddress;
import java.util.Date;
import java.util.HashMap;
import java.util.Map;
import org.apache.storm.task.TopologyContext;
import org.apache.storm.topology.BasicOutputCollector;
import org.apache.storm.topology.OutputFieldsDeclarer;
import org.apache.storm.topology.base.BaseBasicBolt;
import org.apache.storm.tuple.Tuple;
import org.elasticsearch.action.index.IndexResponse;
import org.elasticsearch.client.Client;
import org.elasticsearch.common.settings.Settings;
import org.elasticsearch.common.transport.InetSocketTransportAddress;
import org.elasticsearch.transport.client.PreBuiltTransportClient;
import com.book.domain.AlertEvent;
import com.book.domain.VehicleSensor;
import com.fasterxml.jackson.databind.ObjectMapper;
/**
 * This bolt is used to create index in Elasticsearch.
 *
 * @author SGupta
 *
 */
public class ElasticSearchBolt extends BaseBasicBolt {
    private static final long serialVersionUID = -9123903091990273369L;
    Client client;
    PreBuiltTransportClient preBuiltTransportClient;
    ObjectMapper mapper;String index;
    String type;
    String clusterName;
    String applicationName;
    String host;
    int port;
    public ElasticSearchBolt(String index, String type, String clusterName,
String applicationName, String host,
    int port) {
        this.index = index;
        this.type = type;
        this.clusterName = clusterName;
        this.applicationName = applicationName;
        this.host = host;
        this.port = port;
    }
    @Override
```

13.5 实现用例

```java
  public void prepare(@SuppressWarnings("rawtypes") Map stormConf,
TopologyContext context) {
    mapper = new ObjectMapper();
    Settings settings = Settings.builder().put("cluster.name","my-application").build();
    preBuiltTransportClient = new PreBuiltTransportClient(settings);
    client = preBuiltTransportClient
    .addTransportAddress(new InetSocketTransportAddress(new
InetSocketAddress("localhost", 9300)));
  }
  @Override
  public void cleanup() {
    preBuiltTransportClient.close();
    client.close();
  }
  public void declareOutputFields(OutputFieldsDeclarer declarer) {
  // No further processing is required so no emit from this bolt.
  }
  public void execute(Tuple input, BasicOutputCollector collector) {
    Map<String, Object> value = null;
    // Check type and based on it processing the value
    if (type.equalsIgnoreCase("tdr")) {
      VehicleSensor vehicleSensor = (VehicleSensor)
input.getValueByField("parsedstream");
      // Converting POJO object into Map
      value = convertVehicleSensortoMap(vehicleSensor);
    } else if (type.equalsIgnoreCase("alert")) {
    AlertEvent alertEvent = (AlertEvent)
input.getValueByField("generatedAlertInfo");
    //Converting POJO object into Map
    value = convertVehicleAlerttoMap(alertEvent);
  }
  // Inserting into Elasticsearch
  IndexResponse response = client.prepareIndex(index,
type).setSource(value).get();
  System.out.println(response.status());
  }
  public Map<String, Object> convertVehicleSensortoMap(VehicleSensor
vehicleSensor) {
    System.out.println("Orignal value " + vehicleSensor);
    Map<String, Object> convertedValue = new HashMap<String, Object>();
    Map<String, Object> coords = new HashMap<>();
    convertedValue.put("vehicle_id", vehicleSensor.getVehicleId());
    coords.put("lat", vehicleSensor.getLatitude());
    coords.put("lon", vehicleSensor.getLongitude());
```

```
        convertedValue.put("coords", coords);
        convertedValue.put("speed", vehicleSensor.getSpeed());
        convertedValue.put("timestamp", new
Date(vehicleSensor.getTimeStamp()));
        System.out.println("Converted value " + convertedValue);
        return convertedValue;]
    }
    public Map<String, Object> convertVehicleAlerttoMap(AlertEvent
alertEvent) {
        System.out.println("Orignal value " + alertEvent);
        Map<String, Object> convertedValue = new HashMap<String, Object>();
        Map<String, Object> expected_coords = new HashMap<>();
        Map<String, Object> actual_coords = new HashMap<>();
        convertedValue.put("vehicle_id", alertEvent.getVehicleId());
        expected_coords.put("lat", alertEvent.getStartingLatitude());
        expected_coords.put("lon", alertEvent.getStartingLongitude());
        convertedValue.put("expected_coords", expected_coords);
        actual_coords.put("lat", alertEvent.getActualLatitude());
        actual_coords.put("lon", alertEvent.getActualLongitude());
        convertedValue.put("actual_coords", actual_coords);
        convertedValue.put("expected_distance",
alertEvent.getThresholdDistance());
        convertedValue.put("actual_distance", alertEvent.getActualDistance());
        convertedValue.put("timestamp", new Date(alertEvent.getTimeStamp()));
        System.out.println("Converted value " + convertedValue);
        return convertedValue;
    }
}
```

5. 完全拓扑

拓扑包含所有 bolt 与 spout 的绑定。代码如下：

```
package com.book.processing;
import java.util.UUID;
import org.apache.storm.Config;
import org.apache.storm.LocalCluster;
import org.apache.storm.StormSubmitter;
import org.apache.storm.generated.AlreadyAliveException;
import org.apache.storm.generated.AuthorizationException;
import org.apache.storm.generated.InvalidTopologyException;
import org.apache.storm.kafka.BrokerHosts;
import org.apache.storm.kafka.KafkaSpout;
import org.apache.storm.kafka.SpoutConfig;import
org.apache.storm.kafka.StringScheme;
```

```java
import org.apache.storm.kafka.ZkHosts;
import org.apache.storm.spout.SchemeAsMultiScheme;
import org.apache.storm.topology.TopologyBuilder;
/**
 * This is complete topology to bind spout and all bolts.
 *
 * @author SGupta
 *
 */
public class GeoFencingProcessorTopology {
    public static void main(String[] args) {
        if(args.length <1){
            System.out.println("Please mention deployment mode either local or cluster");
            System.exit(1);
        }
        String deploymentMode = args[0];
        Config config = new Config();
        config.setNumWorkers(3);
        TopologyBuilder topologyBuilder = new TopologyBuilder();
        String zkConnString = "localhost:2181";
        String topicName = "vehicle-data";
        String hcHostName = "localhost";
        String hcPort = "5701";
        String esClusterName = "cluster.name";
        String esApplicationName = "my-application";
        String esHostName = "localhost";
        int esPort = 9300;
        BrokerHosts hosts = new ZkHosts(zkConnString);
        SpoutConfig spoutConfig = new SpoutConfig(hosts, topicName , "/" + topicName, UUID.randomUUID().toString());
        spoutConfig.scheme = new SchemeAsMultiScheme(new StringScheme());
        KafkaSpout kafkaSpout = new KafkaSpout(spoutConfig);
        topologyBuilder.setSpout("spout", kafkaSpout, 1);
        topologyBuilder.setBolt("parser", new ParseBolt(),
1).shuffleGrouping("spout");
        topologyBuilder.setBolt("checkAndAlert", new
CheckDistanceAndAlertBolt(hcHostName, hcPort),
            1).shuffleGrouping("parser","parsedstream");
        topologyBuilder.setBolt("saveTDR", new ElasticSearchBolt("vehicle-tdr",
"tdr",esClusterName,
            esApplicationName,esHostName,
esPort),1).shuffleGrouping("parser","parsedstream");
        topologyBuilder.setBolt("generateAlert", new
GenerateAlertBolt(hcHostName, hcPort),
```

```
        1).shuffleGrouping("checkAndAlert", "alertInfo");
        topologyBuilder.setBolt("saveAlert", new ElasticSearchBolt("vehiclealert",
"alert",esClusterName,
        esApplicationName,esHostName, esPort),
    1).shuffleGrouping("generateAlert", "generatedAlertInfo");
        LocalCluster cluster = new LocalCluster();
        if (deploymentMode.equalsIgnoreCase("local")) {
        System.out.println("Submitting topology on local");
        cluster.submitTopology(topicName, config,
topologyBuilder.createTopology());
        } else {
        try {
          System.out.println("Submitting topology on cluster");
          StormSubmitter.submitTopology(topicName, config,
topologyBuilder.createTopology());
        } catch (AlreadyAliveException | InvalidTopologyException |
AuthorizationException e) {
        e.printStackTrace();
          }
         }
        }
    }
```

13.6 运行用例

在运行代码之前,使用以下命令构建它:

```
mvn clean install
```

这将在 chapter13/target 目录中创建一个 JAR 文件,其名称为 chapter13-0.0.1-SNAPSHOT-jar-with-dependencies.jar。

加载 Hazelcast

执行以下命令将静态值加载到 Hazelcast 中:

```
java -cp target/chapter12-0.0.1-SNAPSHOT-jar-with-dependencies.jar
com.book.simulator.HazelCastLoader <phone_number> vehicle-static-data 10000
```

输出如图 13-6 所示。等待执行下一步,因为程序只读取 Kafka 主题中的最新条目。

Hazelcast UI 将显示一个映射:vehicleAlertInfo,如图 13-7 所示。

图 13-6

图 13-7

1. 生成车辆静态值

执行以下命令以生成车辆静态值，其中包含车辆及由纬度和经度表示的起始位置，并将消息推送给 Kafka：

```
java -cp target/chapter12-0.0.1-SNAPSHOT-jar-with-dependencies.jar
com.book.simulator.VehicleStartPointGenerator 100
```

输出如图 13-8 所示。

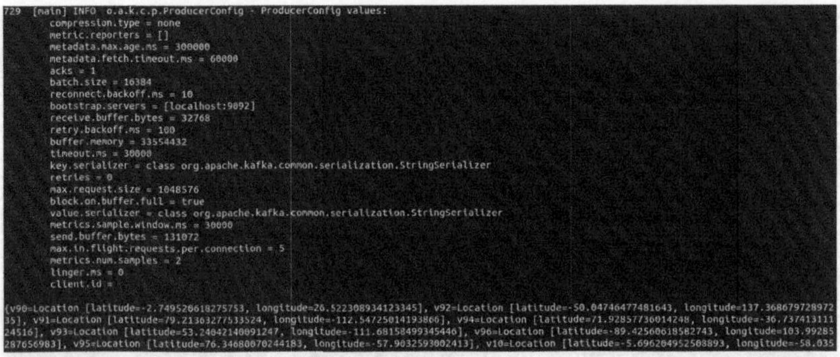

图 13-8

2. 部署拓扑

执行以下命令在 Storm 上部署拓扑：

```
/bin/storm jar chapter12-0.0.1-SNAPSHOT-jar-with-dependencies.jar
com.book.processing.GeoFencingProcessorTopology cluster
```

一旦部署拓扑后，Storm UI 将开始显示**拓扑摘要**，如图 13-9 所示。

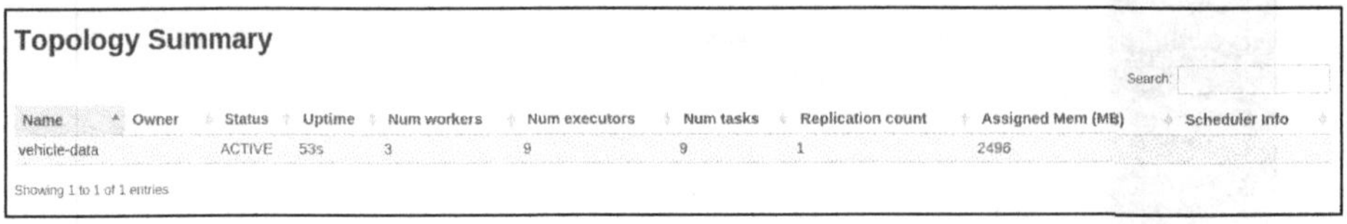
Topology Summary

Search:

Name	Owner	Status	Uptime	Num workers	Num executors	Num tasks	Replication count	Assigned Mem (MB)	Scheduler Info
vehicle-data		ACTIVE	53s	3	9	9	1	2496	

Showing 1 to 1 of 1 entries

图 13-9

单击拓扑名称时，将显示 spout 和 bolt 的详细信息以及其他详细信息，如图 13-10 所示。

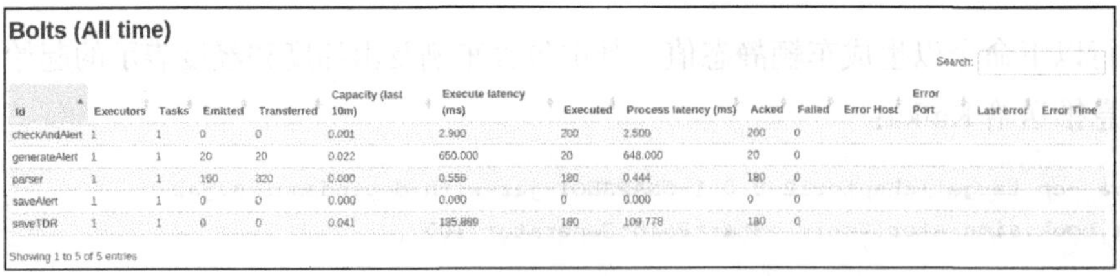
Bolts (All time)

Search:

Id	Executors	Tasks	Emitted	Transferred	Capacity (last 10m)	Execute latency (ms)	Executed	Process latency (ms)	Acked	Failed	Error Host	Error Port	Last error	Error Time
checkAndAlert	1	1	0	0	0.001	2.900	200	2.500	200	0				
generateAlert	1	1	20	20	0.022	650.000	20	648.000	20	0				
parser	1	1	160	320	0.000	0.556	180	0.444	180	0				
saveAlert	1	1	0	0	0.000	0.000	0	0.000	0	0				
saveTDR	1	1	0	0	0.041	135.889	180	109.778	180	0				

Showing 1 to 5 of 5 entries

图 13-10

可以通过单击**可视化**按钮以 DAG 形式可视化整个拓扑，如图 13-11 所示。

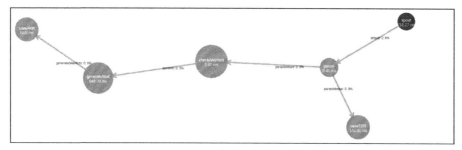

图 13-11

3. 启动模拟器

最后，完成所有设置并启动它们。现在，执行以下命令后，消息将开始推送 Kafka 主题 vehicle-data：

```
java -cp chapter12-0.0.1-SNAPSHOT-jar-with-dependencies.jar
com.book.simulator.VehicleDataGeneration 10000 10000
```

示例输出如图 13-12 所示。

图 13-12

4. 使用 Kibana 可视化

构建具有不同可视化效果的仪表板。首先，必须通过单击 **Management** 添加索引，然后再选定索引模式。在 Kibana 中添加 vehicle-alert 和 vehicle-tdr 作为索引以进行可视化。单击 **Visualize** 后，将显示可用的选项，如图 13-13 所示。

当选择任何类型的可视化时，Kibana 会要求首先选择索引名称，如图 13-14 所示。

现在，根据需要添加可视化效果，以下是几个例子。

- **Tile Map**：通过为 vehicle-alert 索引选择以下选项，Geo 点开始显示在地图上，由于数据点是随机的，因此它们可以位于地图上的任何位置。

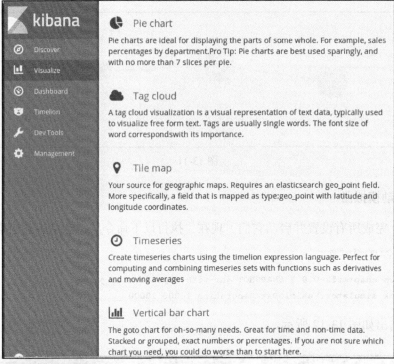

图 13-13

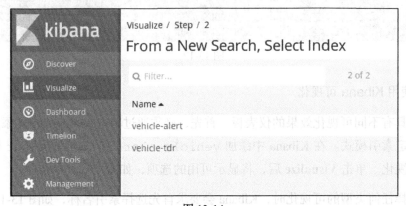

图 13-14

- **Vertical Bar Chart**：此图表显示了所选聚合与 X 轴值的垂直条形图，如图 13-15 所示。

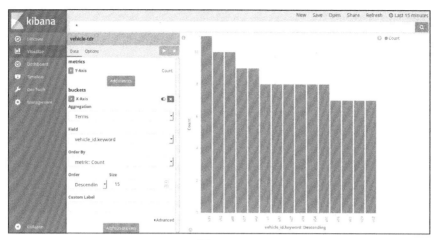

图 13-15

- **Data Table**：获取实际值并执行聚合操作。如图 13-15 所示，按照前面步骤配置可视化以使用它。图 13-16 显示了每辆车的事件总数，图 13-17 是率先发出警报的前 10 辆车的清单。

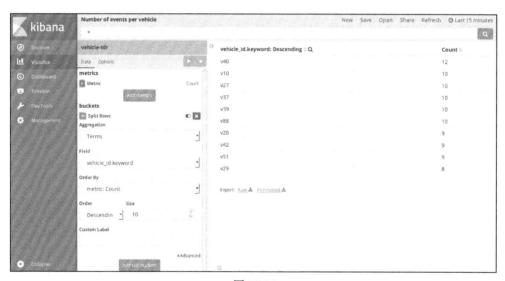

图 13-16

第 13 章　用例研究

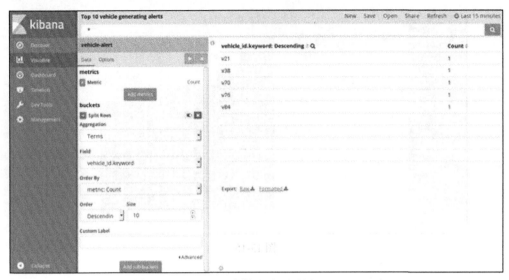

图 13-17

接下来，必须通过选择 **add** 选项添加之前配置的所有可视化效果来构建仪表板。添加所有可视化效果后，仪表板与图 13-18 和图 13-19 类似。

在移动端 WhatsApp 上收到的消息，如图 13-20 所示。

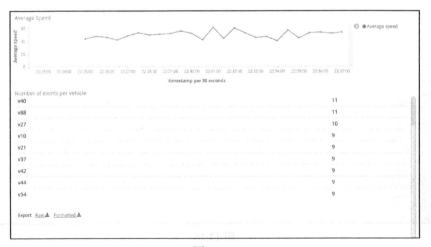

图 13-18

13.7 小结

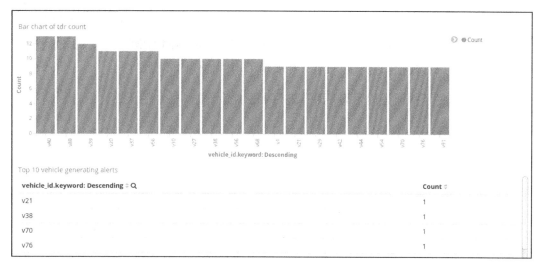

图 13-19

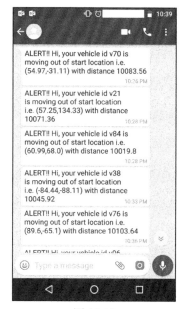

图 13-20

13.7 小结

在本章，我们讨论了用例研究，选择了 Geofencing 用例进行研究。此外，我们解释了与 Geofencing 相关的不同用例；解释了用例实现中使用的数据模型——输入数据集、输出数据集和 Elasticsearch 索引；解释了用于实现用例的不同类型的工具；详细讨论了如何使用不同的工具建立完整的环境，并展示了用例的代码实现。

最后，研究了如何运行用例并对其进行可视化。